Reader Comments

Herb Nordmeyer is one of the industry experts I go to when I need technical assistance. He has the ability to take a complex subject and explain it so anyone can understand it. Last night I started reading a draft copy of ***The Stucco Book—The Basics*** and could not put it down. Besides being chock-full of information, it is written in a humorous style. Have you ever read a construction book that was interesting to read? This one is.

While reading the book, it felt like Herb was sitting in the room talking directly to me as I read the book. I've never had such a special gift before. You can have it also.

Nolan Scheid, MortarSprayer.com

The ***Stucco Book—The Basics*** is written for anyone with an interest in stucco, as well as other cementitious building facades. Whether you are an experienced craftsman or a homeowner with a need to know about stucco, this book is a must-read. Herb Nordmeyer has filled the pages with facts, historic background, and thorough explanations that explain the science and the practicality of concrete, cement, formulas, and stucco or stucco-like products.

Herb presents his 60+ years of experience in a fashion that is chock-full of anecdotes and humorous comments that are relevant everyday life experiences. At some point during this read, you'll stop and wonder: is this a humorous book about a technical subject, or a technical text written with a humorous bent? Of course, the answer is the latter.

Whenever I have a question regarding stucco, masonry veneer, cement, or concrete, I don't think about going to the internet. Instead, I contact Herb; I can be sure to receive an immediate and accurate answer, as well as a full explanation finished with a wry bit of humor. Once you hang up the phone or read the email, you'll wonder how Herb can remember so much information and detail.

Rick Garglano, Eldorado Stone

As I was reviewing ***The Stucco Book—The Basics***, I heard that Andy Rooney had died. I never met Andy, but in reading his writings and listening to him talk, he could make me uncomfortable and at the same time like him all the more, because he was a man who could tell the truth in a more constructive manner than many others. He not only thoroughly knew what he was talking about, but he also cared very deeply. I've known Herb for a number of years and find that, like Andy, he thoroughly knows and cares about any subject he addresses. He is not afraid to poke fun at himself for the mistakes he has made gaining experience and wisdom. If you are a caring professional involved with stucco, this is a must-read book. If you have the good fortune to know or to meet Herb Nordmeyer, you will have no doubt that he not only knows his subject, but he also cares deeply about it.

John H. Koester, Founder and CEO, Masonry Technology

The Stucco Book Series

The Stucco Book—The Basics

Everything you need to know to complete a perfect conventional stucco job.
Published—2012

The Stucco Book—Forensics & Repairs

Everything you need to know to diagnose and fix stucco problems.
Anticipated publication—late 2012

The Stucco Book—Creative Stuccoing

Everything you need to know to implement your creative ideas in stucco.
Anticipated publication—2013

Other Publications by Herb Nordmeyer

We Heard The Wings of Angels

32 devotions by and for cancer patients and their families
Edited by Judy and Herb Nordmeyer
Published—1999

Cancer—An Intense House Guest

A practical guide for living with cancer
By Judy and Herb Nordmeyer
Published—2008

Animals I Have Hated

Humorous animal stories with life lessons
By Herb Nordmeyer
Published—2012

Go to www.NordyBooks.com to find out where to obtain these publications.

Everything you need to know to complete a perfect conventional stucco job.

The Stucco Book
The Basics

By Herb Nordmeyer

Nordmeyer, LLC
Castroville, TX
2012

ISBN: 978-0-9847936-1-7

Library of Congress Control Number: 2011963530

Author portrait by: Yolanda Chapin
Photographic Memories
http://yolandachapin.com

Cover design by: Janice Campbell
The Very Idea
www.theveryidea.biz

Illustrations by: Adam Skoda
Masonry Technology Inc.
www.mtidry.com

Library of Congress Subject Headings: Construction
Plaster
Stucco

Trademarks

Best Masonry is a registered trademark of OldcastleAPG.
Easy Spred is a registered trademark of Peninsular Products, Inc.
Gunite is a registered trademark of Allentown Equipment Co.
Magna Wall is a registered trademark of OldcastleAPG.
Styrofoam is a registered trademark of Dow Chemical Co.
Tyvek is a registered trademark of DuPont.

Published by:
Nordmeyer, LLC
213 County Road 575
Castroville, Texas 78009-2120
www.NordyBooks.com

Dedication

To my wife Judy who has reviewed every draft of this book and all of the other writing that I do. She takes my spelling errors and grammatical errors in stride and drips red ink over them. Even when my word processor cannot figure out what I mean, she soldiers on. In payment, she gets to chuckle at my mistakes and listen to readers gush, “Isn’t he wonderful.”

Acknowledgements

This book, or more correctly, this series of books, has been in the process for a number of years. I intended to write them, but life regularly got in the way.

I've known the Scheid family for a number of years. In the spring of 2011, Sylvia Scheid came on one of the wilderness kayak camping trips that I lead. While on the trip, we celebrated her Quinceañera (fifteenth birthday) in a cave overlooking the Rio Bravo del Norte. After the trip, I took her to the Alamo where I explained how incorrect restoration can damage a building constructed of adobe and soft limestone. She acted like she was thrilled with my explanation, but told me that her father really, really wanted me to get a book written on stucco. She started calling me Abuelo (grandfather), and I started calling her Neita (granddaughter), so what could I do but start writing?

Her father, Nolan, has been looking over my shoulder to ensure that I keep working. On a regular basis he asks questions that require me to think and add another chapter to the series.

Over the years Ricci Woods, currently with Boral Mineral Technology, and I have shared information about stucco and visited sites with problems. On many occasions Ricci has asked questions that challenged me to think.

When I was fourteen years of age, and studying chemistry, my father felt that if I was going to develop cementitious products, I needed to learn how to use a trowel. He was one of the hardest bosses I ever had, because he would never accept anything except the best.

Dr. Rich Klingner asks more questions about cement chemistry than anyone I have ever known and remembers it all so he can ask me about inconsistencies in my statements.

Members of ASTM Committees C 1, C 11, C 12, and C 15 constantly challenged my facts as we developed consensus standards for cements, stuccos, mortars, and masonry units. This taught me that I had to not only have my facts straight, but I had to be able to defend them.

For a number of years I worked with Mary Ortiz. Her detailed knowledge of jobs and problems provided an excellent background upon which to build these books.

When I worked with David Dunaway, he would dredge up problem buildings so I could practice forensic analysis.

Charlie Meador, who operates a stucco supply company in the Dallas area, has worked to ensure that contractors buying from him know the proper way to apply stucco. He has bounced hundreds of ideas off my pointed head trying to ensure that we both understood all of the ramifications of each decision. As a result, there are fewer problem jobs in the Dallas area than in most metropolitan areas.

A number of people have reviewed all or part of this book. Some are experts in the field and some are beginners. I have tried to address each of the questions they had and as a result, this book is much better than it would have been without their assistance.

While there are many others whom I could mention, I will stop with the stucco contractor whose biggest asset was his ten-year-old pickup truck. He sat in my office, admired the 8th grade diploma on my wall (that is another story), and when asked how long he wanted a job to last, responded without hesitation, “Until my check clears.” If I can provide information that reaches at least a few of the contractors who think like him, the effort to write this book will have been worth it. In the meantime, maybe some of the people who hire him will have read this book and will know the questions to ask and the results to demand. Maybe one of them will require that he read this book.

Table Of Contents

Foreword

Herb Nordmeyer became a leading stucco consultant through a lifetime of hard work, constant study, and trying to learn from everyone he ever met. He does not mention that he is a stucco expert. After spending a little time with him, there is no doubt that he is not only a leading stucco expert, but knows several related technologies as well.

His family was in the pozzolan business (see Chapter 1 for a definition), and furnished pozzolans for Falcon Dam, which was the first Bureau of Reclamation dam that used pozzolans. When Herb's father needed a chemistry education, the family cleared off the kitchen table, and Herb and his father studied chemistry three nights a week under the guidance of Herb's mother, a chemist. Herb went on to earn degrees in biology, chemistry, and aquatic ecology, but he considers his real education that home schooling. After following a career based on his formal education for a few years, he got back into the construction materials field.

Herb has spent most of his career as a research scientist with an interest in pozzolans, including fly ash, as well as stucco, lime, and mortar. Unlike most research scientists, he spent much of his time in the field, working with the people who have trowels in their hands, to find out what they needed.

We met Herb when he was organizing the Straw Bale Association of Texas to finish the Greg Tomko Straw Bale Home. Greg was about 10% complete in building his dream home when he fell off the roof and passed away. This left a family, an unfinished home, and inadequate plans, since most of the plans were in Greg's head. Herb and SBAT, along with hundreds of volunteers, finished the home for the family. When we heard about the project, we sent a stucco sprayer to Herb. He responded by writing an operator's manual for us. One of the treasures that came out of the project was the beginning of our friendship with Herb. The Tomko Straw Bale Home serves as a vacation rental/special events center used to support the family (http://www.strawmanor.com).

In the early 1980s, Herb and several associates, formed Rainbow Cement Co., Inc., and produced colored masonry cements and stuccos. In the early 1990s he became associated with Best Masonry & Tool Supply, with facilities in Atlanta, Houston, and San Antonio. His primary job was to develop the greenest line of cement products that had ever been developed. His lab was referred to as "The Skunk Works," after the WWII lab which developed products that most people thought could never be developed. He followed his father's advice:

Just because no one has done it, does not mean it cannot be done, and the man with the trowel must say, "I love it!"

Every year or so, a company would come along and purchase the company that Herb was working for, and he often ended up reporting directly to the company president. Following his retirement in January, 2010, he set up a consulting company, so he is still working for the company president.

People laughed at many of the things that Herb's father proposed. When Herb proposed them thirty years later, they were considered innovative, so Herb has concluded that to be hailed as being innovative, one should listen to one's father. This is typical Herb. He takes his work seriously, but he does not take himself seriously or put on airs.

Herb has been very active in ASTM, serving on Committees C 1 (Cements), C 11 (Gypsum and Related Products including stucco), C 12 (Mortar for Unit Masonry), C 15 (Masonry Units), and C 27 (Precast Products), and had the honor of serving on the Board of Directors of ASTM Committee C 12.

I consider it an honor that Herb asked me to write the forward for this book.

Nolan Scheid, MortarSprayer.com

Preface

My friends tell me I should write a book about stucco. My clients tell me I should write a book about stucco. I look at the internet and find that I am a leading expert on stucco. Seriously, I have been in the construction materials business for most of my life. My father and his father were in similar businesses, and as I was growing up, I listened and remembered when my father talked. This has given me about 90 years of experience to draw on.

Often I do not remember some of these experiences until someone asks a question, then it comes flooding back. An example is the plastering of the ceiling in the Edinburg, Texas, Railroad Depot. It was being redone for use by the Chamber of Commerce. The plaster had little red and blue flecks in it. When these flecks were mentioned to me, I remembered that my father had told me that when he was a teenager, they were building the depot and someone had brought one-eighth-inch pieces of red and blue thread from Reynosa, Tamaulipas, and that they had mixed it in with the final coat of plaster.

There were many techniques that have been used in the past that are not being used today, but they might be an inspiration for someone to develop a new product or a new method of using stucco. If I can get one person to think outside his comfortable little box, this book will be successful.

While I have been in the research and development mode for most of my career, I have considerable experience in each of the following:

- Raw material processing, including pozzolans,
- Laboratory formulation of stuccos, mortars, and other cementitious products,
- Field and laboratory testing of stuccos, mortars, and other cementitious products,

- Blending and packaging of stuccos, mortars, and other bagged cementitious products,
- Application of one-coat and three-coat stuccos in traditional settings,
- Application of stucco in non-traditional settings, such as with straw construction,
- In-plant and in-field quality control and quality assurance,
- Forensic analysis of failed stucco projects, and
- Development of ASTM standards.

There are those who have a deeper knowledge of individual parts of the industry, but I probably have as broad a background as anyone. As part of the R&D, there have been many projects that were discarded. Those that I think have promise, I will mention, that is, if I can dredge them out of my memory. Someone else might have a different angle of approach and make the concepts work.

Chapter 1
What is Stucco?

Before I explain what stucco is and what stucco is not, I want to spend a little time talking about the history of some of the components that go into stucco, and before we get into that, I want to define "concrete." You will find more chemistry in this chapter than you will find in the rest of the book. Do not let this chemistry intimidate you, because I promise I will not use any chemistry that I was not comfortable with by the time I was fifteen years old.

Concrete is a mixture of a cementitious binder and aggregate. Now that I have thrown my favorite word, Cementitious, in so everyone is impressed, in the future I will try to use the term "cement-like." Concrete is a mixture of a cement-like binder and aggregate. If the aggregate is made up of 1 1/2" gravel, ¾" gravel, pea gravel, and sand, the concrete will resemble what most people refer to as concrete, which is used for producing sidewalks and driveways. If the aggregate is sand and the concrete is placed between brick or other building units, we usually refer to it as mortar. If the aggregate is sand and we apply it to vertical walls and to ceilings, we usually refer to it as stucco. You get the idea. When I refer to concrete, I am not always referring to something that is used to build sidewalks.

The Egyptians used gypsum-based concrete. As long as it was used in arid areas, it held up well. They also burned and hydrated lime to make concrete. There are some who maintain that they used pozzolanic technology to form the blocks that made up the pyramids, but that theory no longer seems as plausible as it did a few years ago. Where would they have mined that much pozzolan? You have to wait for another page to find a definition of pozzolan.

The Greeks made their concrete with a lime binder. The lime got hard by drying and by combining with carbon dioxide from the air to form calcium carbonate. That is what limestone is made from. At some point they discovered that if they mixed volcanic ash with the lime putty/aggregate mix, the resulting products would get harder and be more durable. The Romans applied their engineering skills to the Greek concept, and concrete made from volcanic ash, lime putty, and aggregate became a very popular building material in what is now Italy.

Limestone is calcium carbonate or calcium-magnesium carbonate. Lime putty is formed by burning limestone (about 1,650° Fahrenheit) to form quick lime (calcium oxide or calcium-magnesium oxide). In the process, carbon dioxide is given off. The quick lime is then hydrated. Hydrating is adding enough water to chemically combine with the calcium oxide to form calcium hydroxide. In the past, to hydrate it, a trench would be dug about 18″ wide by about 3′ to 4′ deep. Quick lime would be slurried with water (the water boiled, and sometimes chunks of quick lime would explode, so it was not a very pleasant process) and then would be buried in the trench for three months or longer. It would be dug up, mixed with other ingredients, and used. Probably the first limestone was burned on top of the ground in a cooking fire and then inadvertently was found to make a good building material. Since cooking fires did not produce enough quick lime to meet the demand, lime-burning pits were developed. These were pits that were dug at the edge of a quarry, or along a river bed, so they could be loaded from the top, and a fire added through a tunnel from the bottom. In time, above-ground kilns were built. It was not until the 20th century that rotary kilns were used to burn lime, and hydrators were used to hydrate the quick lime.

Volcanic ash is fine particles that are given off by volcanoes. The finer the particles, the more active they are. Chemically they are predominantly amorphous (non-crystalline) alumino-silicates, which is another term for glass. Volcanic ash can be found in primary deposits (where it fell), or it can be in secondary deposits (where it has been blown or was water-washed). It can be consolidated to form volcanic tuff. The term pozzolan developed because the original volcanic ash the Romans used came from near the town of Pozzuoli near Mt. Vesuvius.

When volcanic ash is reacted with lime putty in the right concentrations, hydrated monocalcium alumino silicates and dicalcium alumino-silicates are formed. This is a slow process and is endothermic. That means that the reaction absorbs heat. Since heat is required, the reaction is faster in a warmer climate than in a cold climate. The reaction stops when the temperature is about 40° Fahrenheit.

Portland cement is formed by adding limestone, clay, and a few other materials together in a kiln and heating them until they melt. In the process, carbon dioxide is given off and calcium oxide is formed. The calcium oxide and the alumino silicates from the clay melt and react chemically. Mono-calcium alumino silicates, dicalcium alumino silicates, and tricalcium alumino silicates are formed. This is actually an over-simplification, since there are hundreds of compounds that are formed. The molten cement is then cooled to form cement clinker. The cement clinker is then milled to produce Portland cement.

When Portland cement hydrates, a calcium oxide ion is released and immediately combines with a water molecule to form calcium hydroxide. Example:

Tricalcium alumino silicate + water
Yields
Dicalcium alumino silicate hydrate + calcium hydroxide

The dicalcium alumino silicate hydrate is the same compound that the Romans were producing by reacting volcanic ash and lime putty, but the Romans controlled things so there was not any extra calcium hydroxide in their final product. While many will deny it, the unreacted calcium hydroxide leads to degradation of the concrete (stucco, mortar) over time. The Roman system produced a better concrete, and we can see evidence of their work after 2,000 years. The problem with the Roman concrete was that the setting time was longer than our fast-paced life will allow.

When the Romans did not have volcanic ash, they found they could under-burn brick and crush them and have similar results. Modern metakaolin is produced in a similar fashion.

The Iliad speaks of making floor tile. The tile were formed and then cured by using them as roof tile for a few years. After that they were used as floor tile. These floor tiles were made using the pozzolanic technique.

Even though historians, as recently as late 20th century, told us that pozzolanic technology was lost with the fall of the Roman Empire, and that it is impossible to re-create the technology, it has been around. Some historians just did not know where to look. Santa Sophia was built in Turkey in the 5th century using pozzolanic technology, and about a thousand years later the Kremlin in Moscow was built with pozzolanic technology. My father and I both spent most of our lives working with the technology.

While the pozzolanic technology produced excellent products, in cool weather it was slow to set. When the temperature approached 40° Fahrenheit, the reaction came to a standstill. Often in more northern climates, hydrated lime mortars and stuccos would be used rather than pozzolanic mortars and stuccos. In some areas there were natural stone that could be burned and ground and would, when water was added, hydrate and form stones again. These were called natural cements. Numerous people worked on combinations of soils and rocks to burn together to form a replacement for natural cement. Apparently at least five different people came up with a workable product during the first half of the 19th century, and, depending on which country you study history in, you will find that different people created the first Portland cement. The name Portland cement comes from the fact that the product that was first produced in the south of England, looked like Portland Stone when made into concrete. Portland Stone was a building material that was quarried in the south of England. The first

Portland cement was produced in the US in about the 1870s. Shortly thereafter small amounts of Portland cement were added to lime putty to accelerate the set of mortar. Then it was used to produce stucco.

After all of that introduction, we get to ask the question, "What is stucco?" ASTM (American Society of Testing and Materials) C 926 (Standard Specification for Application of Portland Cement-Based Plaster) in section 3.2.23 defines stucco as "*Portland cement-based plaster used on exterior locations.*" If we use ASTM's definition of stucco, then there was no stucco produced prior to the invention of Portland cement, even though the material applied by the Romans to walls is chemically essentially the same as what we now call Portland cement-based plaster. A plaster is a paste that is applied to a wall to protect the wall or to enhance the aesthetics of the wall.

Wikipedia has a much broader definition of stucco. It defines stucco as "a material made of an aggregate, a binder and water. Stucco is applied wet and hardens to a very dense solid. It is used as a coating for walls and ceilings and for decoration. Stucco may be used to cover less visually appealing construction materials such as concrete, cinder block, or clay brick and adobe." Wikipedia's definition includes hydrated lime plasters and does not limit the definition to exterior applications.

So, what is stucco? It is somewhere between a Portland cement-based plaster used in an exterior location, and something you put on the wall and hope it stays there.

Now we will add two more terms to the mix: shotcrete and Gunite.

Both terms are used for a stucco-like material that is sprayed onto a surface. Numerous people use the terms interchangeably. Gunite, oftentimes spelled gunnite, is the registered trademark of the Allentown Equipment Company and should only be used in conjunction with their dry process shotcrete equipment. The dry process was invented at the beginning of the 20th century and Gunite was a brand name. The term shotcrete was developed in the 1930s. It was not until the 1950s that the wet system was developed. Shotcrete is usually used where structural strength is required, such as building swimming pools, stabilizing highway embankments, and stabilizing mining tunnels. Often the concrete is blown into a framework of rebar.

Ferrocement is a technique of applying a stucco mix to multiple layers of a steel mesh, such as chicken wire or stucco netting. It was used during WW II to build ships. Back then the stucco mix had to be very dry and hard to work with, since the water/cement ratio had to be as close as possible to 0.35 for the stucco to gain the strength that was needed. With modern water-reducing agents, a 0.35 water/cement ratio can be obtained with a much more fluid mix. With the development of fibers, especially the finer fibers, more and more people are adding a higher concentration of fibers to stucco and plastering around armatures to produce sculptures.

When a plaster mix is used as a façade, it is usually referred to as stucco, whether it is applied with a trowel, blown on with a pump and gun, or thrown onto the wall with a shovel. Even though some builders and architects seem to think that stucco should hold shoddy design and workmanship together, stucco is not designed to be a structural material. Shotcrete can be referred to as structural stucco. For the most part, we will not be concerned with shotcrete, or ferrocement, in this book, but you will learn some techniques so you can produce self-supporting structural stucco.

Editor's Note: In the construction industry, some words are used as both singular and plural. This includes brick, flagstone, stone, and tile. They are used in that manner in this book.

Chapter 2
Making Stucco the Old Way

My grandfather, Edward F. Nordmeyer, was highly educated. He made it through part of four years of school, and then he found that he could learn whatever he wanted from books and from experimentation. During his lifetime he did many things, including becoming a civil engineer and a surveyor. He was the engineer for the McAllen, Texas, storm sewer project in 1932-1933. My father served as his assistant. They burned their own limestone in a pit called a *calero*. The quick lime was hauled about 15 miles to the City of McAllen in horse-drawn wagons and then slurried and buried in a trench to made lime putty. They leased a brick plant and made their own brick. A trench up to 27 1/2′ deep was dug by hand, and then a double-walled brick tunnel, 5 1/2′ inside diameter, was laid. It was plastered on the inside with a hydrated lime plaster. The plaster was made from about 6 shovels of hydrated lime, 1 shovel of Portland cement, and approximately 20 to 24 shovels of sand. In the late 1990s I examined the storm sewer. The internal plaster was gone, but the brick and the mortar between the brick were still there.

I asked my father how he learned so much about concrete and other cement products. He laughed and said it all started with Concrete Charlie. Concrete Charlie was a blind sidewalk contractor in McAllen. I never learned his last name or how he got into the business. My father said he was Danish. When my father was about eight years old at the end of the Great War (the name World War I was not used until we were in World War II), Concrete Charlie's crew was pouring a sidewalk near where my father lived. Being a boy, my father had to carve his name in the sidewalk. He got caught. He was given the option of working for free to pay for the damage he had caused, or Concrete Charlie would talk to my father's father. My father elected to work. In the process he found that there were numerous things to learn about concrete and that Concrete Charlie was an excellent teacher. Before long my father took ownership in the sidewalks he helped construct, and was spreading the word that no one had better carve any initials in any of his sidewalks.

Expanded metal lath was invented in the late 1700s but did not come into use in the United States until the early part of the 20th century. Prior to that, stucco work was applied to thin boards that were about 1/4″ thick and spaced with a gap between them. In cases of fire, the lath strips would feed the fire, so the early uses of expanded metal lath were to build walls that were more fire-resistant. An interesting fact about the wood lath, which was still commonly used up until WW II, was that when the scratch coat was applied, the wood would absorb some of the moisture from the stucco and swell. As the wall cured, that moisture was transferred back to the stucco resulting in a better cure and higher-strength stucco. A downside of this was that as the wood lath dried, the wood would shrink and break some of the bonds holding the stucco to the wood lath. But this was a mixed blessing. The wood lath was nominally 1″ or 1 1/2″ wide. Expansion and contraction was not much. The length of the lath strips, which spanned from stud to stud, did not change much, so there was little cracking caused by the expansion and contracting of the wood. When expanded metal lath came into common usage, it was often backed by plywood (since then, OSB is often used rather than plywood, since it is cheaper) (OSB stands for Oriented Strand Board. It is a panel that is made from wood chips that have been glued together under pressure. Commonly the installers failed to gap between the sheets, and moisture expansion would cause the sheets to buckle between the studs. Suddenly a new method of failure of stucco was born. My grandfather did not have to deal with it, since he seldom used wood-based sheathing.

Sometimes there was not any lumber to build walls or expanded metal lath to apply the stucco to, so a common method of building partition walls in the Lower Rio Grande Valley of Texas at that time was to hang a sheet of burlap and firmly attach the top and the bottom. Then a plasterer would be placed on each side of the wall and using their trowels in unison, would apply a thin coat of stucco or hydrated lime plaster to each side of the burlap. After the first coat had dried, a second coat would be applied to each side. Depending on the use of the wall, a third coat might be added. This resulted in a wall that ranged from about 1″ thick to about 2″ thick. If the mix was a hydrated lime plaster, they could add Portland cement to speed the set and make it harder. If they did not have Portland cement, they could accomplish the same thing by adding a little molasses. If the mix contained Portland cement, this was not done, since the sugar in the molasses would kill the cement-setting reaction. In the 1990s my father and I visited several stores in Weslaco that he had helped build, and the partition walls were built in this manner. The people working in the stores had no idea how the walls were built.

They also had an interesting way to build a concrete roof. A temporary framework would be installed; expanded metal lath would be placed over the framework and nailed as little as possible. A thin layer of a stucco mix with pea gravel would be shoveled over the lath and allowed to cure. Reinforcing would be added, and then concrete with larger aggregate would be hauled to the roof with buckets and poured on the stucco. After the concrete was cured, the temporary framework would be removed. Rather than being perfectly flat, these roofs had a slight dome to them to help carry the weight of the roof to the walls.

There was much irrigation in the Lower Rio Grande Valley, and water was being lost to seepage. Valley Brick and Tile Company in Mission developed a canal-lining tile with help from my father. He then oversaw the installation of tile in many miles of canals. The process was to dig and form the canal, install the tile with mortar between them, and then plaster the inside with stucco.

Concrete ships were built during WWII. A framework of reinforcing steel and wire mesh would be built, and a plaster would be placed inside and out to press the stucco into the mesh. Since high range water reducers had not been invented at that time, and there was a desire to keep the water/cement ratio down to about 40% to gain the greatest strength, the mix was very stiff and required a great deal of effort to press the stucco into place. The burlap walls mentioned earlier did not require as stiff a mix, so were much easier to build.

During the 1930s a company called Holiday Hill Stone started using volcanic ash, mined from the Starr County area of South Texas in their Lower Rio Grande Valley plant where they made masonry units from a concrete mix and cured them using an autoclave. This resulted in some experimentation of mixing the volcanic ash with hydrated lime stucco, without resorting to the addition of Portland cement. Portland cement would speed the time of set. When the volcanic ash was used, the time of set was not improved, but the stucco did grow harder.

Then our lives changed drastically. When Portland cement hydrates, heat is given off. This heat raises the temperature of the concrete and causes it to expand. At the same time the outer skin of concrete is being cooled by the ambient air and is shrinking. This causes stress on the outer skin of the concrete and cracks occur. Once a crack had occurred, it would continue to grow as long as the temperature inside the concrete pour was rising and the temperature of the skin was falling. This was not a major problem with small pours such as building foundations, but for a large dam it was a major problem. A crack would allow water and oxygen into the concrete and could cause deterioration of the steel reinforcing in the concrete, or it could allow water to flow through the concrete. This problem could be countered by either making small pours and letting them cure before the next pour, or by installing miles of copper tubing in the concrete and pumping cooling water through the tubing. Neither was a good alternative, and the Bureau of Reclamation was looking for a better alternative. In the 1930s and 1940s, Dr. Raymond E. Davis with the University of California at Berkeley developed an alternative. He convinced the Bureau to change the way they built dams. He stated that if a pozzolan were added to the concrete mix, it would react with the calcium hydroxide given off by the Portland cement as it hydrated and produce more cement binder. Further, the reaction was an endothermic reaction, so it would cool the cement as the reaction took place. If the concept worked, a stronger concrete could be produced using much larger pours and without all the miles of copper tubing. A better product could be produced at less cost. Tests at the Bureau's lab in Denver confirmed that it would work, and Falcon Dam on the Texas-Mexico border was selected as the first dam to be built with this innovative method. One of the reasons this dam was selected was because there were volcanic ash deposits in the area. The Bureau went looking for a pigeon (independent contractor) to build a plant, mine the volcanic ash, process the ash, bag it, and ship it to the dam site. Valley Brick and Tile, headquartered in Mission, Texas, and with a plant near a volcanic ash deposit near Rio Grande City, Texas, was their pigeon. VB&T saw it as an opportunity, and a contract was signed. The Bureau had developed specifications that had to be met. Soon, Pozzolana, Inc., was in business, and shortly they were having problems meeting the specifications. The Bureau was considering pulling their performance bond. A relative owned a portion of VB&T, and since my father had helped them with the development of the canal-lining tile business and a structural roofing tile business, as well as having helped them solve other problems, he was asked to step in and solve the pozzolan plant's problems. He did and eventually became Director of Research and Development for Pozzolana, Inc., and Rio Clay Products, but that was after the immediate problems were solved.

The plant ran twenty-four hours per day, and a federal government inspector was on site to ensure that all production was in spec. One night, when a Cold Norther (This is Texanese for a storm out of the north that requires a jacket.) was blowing, my father and I stopped past the plant. We went into the tool shed where the inspector was holed up trying to stay warm and talked to him for a few minutes. Paul Barrett, who was the father-in-law of one of the owners and also was in charge of the night shift, walked in and picked up a 12-gauge shotgun from behind the door. My father asked, "Paul, got labor problems again?" Paul said, "Yes," and walked out. A few minutes later we heard the shotgun, then after a minute or so a second blast, and then a third blast. Paul walked back into the tool shed, put the shotgun behind the door, and said, "They're taken care of."

When the rotary kiln would get too hot, the volcanic ash would start to melt and would form "globs" and stick to the wall of the kiln. Even if the fire was lowered, the "globs" would remain and lead to a blockage of the kiln. Then the kiln would have to be shut down, and a man with a single jack and a star drill would have to mine the mass out. We had learned that when the fire was lowered, the "globs" could be shot out of the kiln with a shotgun. We thought nothing more about the incident, but three days later, federal investigators showed up to investigate our labor practices and interviewed all of our employees. Apparently federal inspectors do not have a sense of humor.

With the completion of Falcon Dam, we started looking for other ways to use the processed volcanic ash and soon were selling a bag of pozzolan and a bag of hydrated lime to use as a mortar or stucco. The masons and plasterers loved the feel of the product on the trowel, and then we had a problem. While visiting a job site where the walls of a brick building were up to window ledge height, the contractor stated that he knew that hydrated lime did not do anything for a mortar, so he had eliminated it. That was the day I learned how hard it was to convince a contractor not to use a shortcut he had discovered, that he thought would make him lots of money.

You already know the chemistry for the pozzolanic reaction. A pozzolan is very slightly soluble in water. Hydrated lime is very slightly soluble in water. When the two products are mixed at a ratio of about 3 parts pozzolan and 1 part hydrated lime (activity of the pozzolan and the fineness of the particles impacts the exact ratio), the soluble components chemically combine to form hydrated calcium alumino-silicates. This allows more of the pozzolan and more hydrated lime to eventually be dissolved and then to combine. The reaction continues until the mixture dries out, or until there is no more of one of the components to dissolve. Since the pozzolans are often spherical in shape, and since the hydrated lime tends to be slick, the mix has great workability. Water is not sucked into most pozzolans, and the water demand for Type S hydrated lime is quickly satisfied with five minutes of mechanical mixing, so the mix has a great board life. There are places in Mexico where the soil is of volcanic origin, and the natives just mix water and hydrated lime with the soil to make stucco.

When I was building a straw home, my father, who was in his early 80s, came out to help. One day we needed to stucco several small areas. By this time, I had a national reputation as an acknowledged expert concerning stucco. It was too small of a job to fire up the mortar mixer, so I mixed the first batch in a wheelbarrow. All components are added to the wheelbarrow dry and are dry-mixed. Traditionally, each slice of the mixing hoe should take about a half inch layer of material and move it to the near end of the wheelbarrow. It is moved from end to end in the wheelbarrow until the mix looks uniform. It is then pulled back and forth twice and water is added and the process continues. My father picked up a hawk (a square metal sheet with a handle on the bottom to hold stucco) and trowel and started to apply the stucco. After the second pass, he turned to me and asked where I had learned to mix stucco. "From you." I replied. "You didn't learn to make this sorry mud from me," he retorted. "I'll mix the next batch and let you apply it." While he mixed the next batch, I applied the first batch. He used the same components and in the same proportions as I had. When the second batch was ready, I started to apply it, but it was creamier than mine. I let a little of it slip off my hawk and hit the ground. He said, "You can't even apply stucco. You go find something else to do, and I will do the mixing and the application. I never met anyone under 75 that had sense enough to mix stucco in a wheelbarrow." He had been taking 3/8″ bites of the components to mix them rather than the 1/2″ bites that I had been taking. The difference was quite apparent. I went off and found something else to work on and he finished the little stucco job. Also, even though I had taught many people to use the 1/2″ bites with the trowel, I changed to teaching using 3/8″ bites.

Chapter 3
Three-Coat Stucco

According to John Maddox Roberts, it is reputed that in the year 703 (by Roman reckoning about 53 BC) Decius, a plebeian aedile (literally—working class official—possibly a building inspector) stated: **"In Rome, honest building contractors are as common as volunteer miners in the Sicilian sulfur pits."**

By the above quote, I am not implying that all building contractors are dishonest. People will convince themselves that a cheaper way of doing something is just as good as a more expensive way of doing something. Sometimes that is the case, and that is where improvements in the industry come from. At other times, the cheaper way is not better, and this is where shortcuts get a bad name. In the stucco field, we have a plethora of both.

There must be a stable structure to hold the stucco. Stucco is a façade and is not designed to hold faulty workmanship together, even though this has been tried repeatedly by architects and contractors. Lawsuits have regularly been filed against stucco suppliers and stucco applicators, when the stucco does not compensate for a faulty foundation or faulty structural framing. That having been said, if the stable structure is in place and three-coat stucco is applied to it, it should provide an excellent façade with only minor cracking.

According to the International Building Code, and most other codes, stucco is a 3-coat process consisting of a scratch coat that is nominally 3/8″ thick, a brown coat that is nominally 3/8″ thick, and a finish coat that is nominally 3/8″ thick, for a total thickness of 7/8″. This stucco is applied to expanded metal or other lath. If the stucco is applied to concrete, masonry, or similar substrates, it can be applied in two coats and can be thinner. Stucco, by definition of the code, has a 1-hour fire rating. Most codes require a lath inspection, but do not require an inspection of the stucco.

About twenty years ago, Paul Taylor, who was my assistant at the time, and I discussed whether the City of San Antonio had a minimum thickness for stucco. Paul's wife ended up calling the San Antonio building inspector's office and asking if there was a minimum thickness for stucco. They could not find one. After a little further investigation, we concluded that a stucco wall in San Antonio had to have lath on it but did not have to have any stucco on the lath. During the lath inspection, the inspector looks at the thickness of the casing bead and weep screed, to ensure that with conventional stucco walls, the thickness is three-fourths-of -an-inch.

Most plasterers know how to apply three-coat stucco. That is not to say that they always follow the best procedures, but they normally do a reasonable job. However, when it comes to one-coat stucco, which is covered in the next chapter, or EIFS, which is covered in the chapter after that, many plasterers do not have a clue as to how to apply it without incorporating problems into the system. There are separate chapters in this book on lath, stucco formulae, and other aspects of applying stucco. Some of these are applicable to one-coat stucco, and when they are, this will be mentioned in the one-coat chapter.

There are some things regarding stucco installation that are pretty well universal. One of them is to allow drainage; water gets behind the façade; if it does not get out, problems occur. A second is to not bury the lower portion of the stucco into the ground. This is especially true in areas that have termites, because termites can easily enter a building that has stucco touching the ground, without ever leaving telltale termite mud tunnel. Windows and other penetrations need to be flashed and sealed, so that water does not get in. If cracks occur in the stucco, water could get behind this stucco during rainstorms. If it cannot readily get out, problems can occur. One-coat and three-coat stuccos should never be considered as waterproof.

We will cover weather-resistive barriers (WRB), or as some people call them water-resistive barriers, in a future chapter.

When it comes to lath and lath accessories, we will cover them in a future chapter.

Three-coat stucco can be purchased as a sanded mix in a bag, in a bulk bag, or in a silo. It can be purchased as a concentrate in a bag, in a bulk bag, or in a silo. It can be mixed from Portland cement, hydrated lime, and sand. Each of the delivery systems has its advantages and disadvantages. We'll start with the biggest one: silos. Silos can be loaded with a sanded mix, or it can be a two-compartment silo with one compartment loaded with sand and the other compartment loaded with concentrate. A metering system feeds into a mixer. Often this is an automated screw-type mixer that can mix a wheelbarrow full of mud or however much is needed. The silos can be hauled out from the blending plant loaded, or they can be filled at the site using bulk bags. Once on the site, the silo system is very convenient, especially in a residential development, where several houses are being built adjacent to one another. A problem is that whether the sand is premixed with the concentrate and in the same compartment of the silo, or in a separate compartment of the silo, it needs to be kiln-dried. Moist sand in a silo tends to pack, bridge, and flow poorly from the silo. Kiln-dried sand tends to hold onto air bubbles when water is added; this leads to increased air content in the mortar and lowers the ultimate compressive strength of the mix. If the mix is robust, there is no problem, but if the mix is marginal to begin with when tested in the lab under ideal conditions, the mix may be weaker coming out of the silo. An additional problem is that drying sand and then moistening it is not as "green" (environmentally friendly) as using moist sand.

Stuccos are traditionally packaged in 1 ft^3 bags, which nominally weigh about 80 pounds. Sand is traditionally delivered to a job site in loose, damp condition. A cubic foot of such sand will contain about 80 pounds of sand and about 6 pounds of water. Three ft^3, 4 ft^3, or sometimes 5 ft^3 of sand are mixed with 1 ft^3 of stucco concentrate. The amount of sand used varies with the formula of the concentrate, the particle-size distribution of the sand, the substrate the stucco is applied to, and the skill of the applicator.

If the same 1 ft^3 bag is used, and it is filled with a sanded mix, the bag weighs 100 to 105 pounds. Some people wonder how you can put more sanded mix than concentrated mix in a bag. Picture a quart jar filled with marbles. It will hold a definite number of marbles, and you cannot pack them in any tighter to get more marbles in. But you can take sand and fit in between those marbles. Whether we are dealing with particles that are the size of marbles or the size of sand grains or the size of Portland cement grains by having a variation in particle size, the mix can be denser. Generally the denser a mix can be made, other things being equal, the stronger the mix will be.

Bags are a convenient way to handle stucco. They can be palletized and handled with a forklift, but each bag is light enough where a man can pitch it onto the mixer deck without help. Most mixer decks are equipped with a cutting tool to slice open the bags. With paper bags there is always waste paper that needs to be disposed of. Dumping the paper bags is a dusty operation, not only for the operator who is pitching the bags onto the mixer deck and pulling off the empty paper bags, but also for anybody downwind.

Masonry cement, either sanded or concentrate, is often used as stucco. There are some people who use masonry cement plus one shovel of Portland cement as their stucco concentrate. While masonry cement meets the *ASTM C 926* standard, there are some masonry cements which do not provide the durability of other masonry cements. Those masonry cements with a high volume of entrained air are a pleasure to trowel, but they often are not as durable.

Portland cement and hydrated lime, blended on the job site or pre-blended in bag, have been used as stucco for over 130 years in the United States.

Now that the components are on-site, they need to be combined. If a silo system is being used, it needs to be hooked up to a water source and an electrical source. Often silo systems have a computer system that pulls the correct amount of concentrate, the correct amount of sand, and adds the correct amount of water. The materials are dumped into a paddle mixer or are fed to a screw mixer. To get a wheelbarrow full of stucco is simply a matter of turning a switch to start pulling material out of the silo and flipping the switch off when an appropriate amount of material is pulled, or programming the computer to deliver a specific-size batch and just poking a button to start the process.

If a mechanical mixer is being used, the process is a little more complicated. An estimate is made as to how much water will be needed, then about 1/4 of that amount of water is added to the mixer. Half of the sand is added. Traditionally, sand is measured in shovels. The traditional shovel used is a number two square shovel. There are people who use cubic-foot boxes to measure the sand; there are others who use 5-gallon buckets to measure the sand. Four 5-gallon buckets hold very close to 3 ft^3. There is no way that some of the people from academia and architectural firms will ever believe that a shovel can be calibrated more accurately than the 1 ft^3 box. It is not only possible, but it happens periodically. As mentioned earlier, 1 ft^3 of loose, damp sand contains 80 pounds of sand. If you were to dry that sand, you would find that you can put about 100 pounds of sand in a 1 ft^3 container. So if the sand is dry, and your mix calls for 3 ft^3 of loose damp sand—240 pounds of dry sand—you could easily put 300 pounds of sand in the mix while trying to be very precise. With a shovel the operator can lose count and rather than putting in 20 shovels, he may put in 22 shovels of sand, or maybe 18 shovels. But if he is an experienced mixer operator, you will notice that he periodically looks into the mixer. His finished batches all fill the mixer to the same level. This means that since he is using the same amount of concentrate, he is also using the same amount of sand in each mix, whether that sand is very dry, at the optimum moisture content, or is very wet. If on the other hand the mixer operator is new on the job, the sand/cement ratio may vary considerably.

When starting a mix, after the water is added to the mixer, about half of the sand needs to be added. If admixtures, fibers, or pigment are to be added at the job site, they need to be added next. The sand/water mix acts as a mill and breaks the admixtures and pigments into individual particles, and breaks the fiber bundles into individual fibers. These components work much better when they are broken down in this manner.

Then the cement portion of the mix should be added. As this is being added, the operator needs to watch the consistency of the mud. Mud is commonly used to refer to the wet stucco from the time it is in the mixer until it is smeared on the wall. If it starts thickening up, he may need to add more water to keep the mud very mixable. If it gets too stiff, it can stop the mixer. Then the remainder of the sand is added one shovel at a time. As it is being added, more water may need to be added to keep the mud mixable. After all components have been in the mixer and they have mixed for about three minutes, the operator needs to adjust the final fluidity of the mix. This is done by stopping the mixer and taking a sample, or by dumping a small amount of mud onto a hawk. If the stucco will stay on a trowel when the trowel is held horizontal with the handle below the blade or with the blade at a 30° angle from horizontal, but will slide off the trowel when it is held steeper than 45° from horizontal, the mix is about right. Most experienced operators can look at the way the mud turns in the mixer and accurately judge whether it is an optimum mix. Do not ever put your hand, trowel, or a cup on the end of a stick into the mixer to obtain a sample while the mixer is turning. The mixer blades can grab the hand or the stick and do damage.

If the stucco has not been mixed for about 5 minutes, it may absorb some water and get too stiff after it is dumped in a wheelbarrow. To overcome this stiffness, additional water can be added. This is known as retempering. If you retemper after the cement particles start to chemically hydrate, bonds that lead to compressive and tensile strength will be broken, and the stucco will not develop the strength that it could potentially develop. Normally you are safe if you don't retemper more than once and you do not retemper after one hour has passed. If the stucco is mixed well for 5 minutes, retempering normally is not required. Mortar for laying brick and other unit masonry is much more likely to have to be retempered since it is used at a much slower rate, and we see degradation based on that retempering. One of the worst cases of retempering that I ever saw occurred in San Antonio, Texas. The stucco on several houses was weak, so I showed up unannounced for an inspection. The man running the mixer mixed up 4 batches of stucco. Each batch consisted of two bags of stucco concentrate and about 6 ft^3 of sand. That means he had mixed up approximately 24 ft^3 of stucco. They didn't have enough wheelbarrows to hold all of the mud, so he stockpiled mud on sheets of sheet rock. Then he joined the other plasterer on the job, and they applied the brown coat. The area they were working had a number of penetrations, such as windows, so they could not stucco very fast. I was sitting in my pickup across the street reading a book, so they didn't realize they were being inspected. The stucco had been mixed between 8:00 and 8:30 in the morning; they finished it up about noon. They washed up, ate lunch, and the mixer operator mixed up three batches of stucco for use during the afternoon. I departed and reported to the man who had hired me, and he assured me that things like that could not happen on one of his jobs. The next day he went out to look, just to prove that I was wrong. No wonder the stucco was soft and crumbly several days after it had been applied.

You need to take the stucco from the wheelbarrow or a mortar board and get it on the wall. This is done by picking the stucco up, placing it on a hawk, placing it on the trowel, and applying it to the lath. I like to start low on the wall and work up. The next trowel-full is started where the previous trowel-full left off and continues up the wall. Others like to start high on the wall and work their way down. They state that this prevents spillage from messing up their newly troweled surfaces. If the trowel is moved in a downward direction, there will be more of a tendency for the stucco to fall off the wall. When applying the stucco to the wall, the trowel should be held so that the leading edge is a little ways out from the lath. Pressure should be applied so that the stucco fully penetrates the lath. When lath is attached to the wall starting at the bottom and working up, with each sheet overlapping the sheet below, it is possible for the trowel to catch the bottom edge of a sheet of lath, and cause the lath to bend and curl. As a result, some lathers start at the top and work their way down when installing the lath. As long as the weather-resistant barrier is lapped correctly, such a system works; and the man with the trowel in his hands does not have to worry about catching a bit of the lath.

When finishing the sweep of the trowel, I like to add a little twist. This is done to help break the bond between the face of the trowel and the stucco. In the process, the leading edge of the trowel is brought further away from the wall. If you were simply to stop moving the trowel and pull it away from the wall, stucco would be pulled off the lath.

Everybody knows that the first coat of stucco is called scratch coat, and it's called scratch coat because a tool is used to impart scratches on the surface. On a vertical wall, the scratches are in a horizontal direction. The concept is to increase the mechanical bond between the first coat of stucco and the second coat of stucco that is applied. Twenty years ago I did some testing, but I never wrote it up and got it published. I made several panels. Some I scratched in the conventional manner. Some I used the edge of the trowel to drag over the surface of the stucco to level it and roughen it. This is called back-dragging. Some were left with a rather slick surface. After the panels had cured for a day, I used the trowel and knocked the crumbs of stucco off of half of the scratched panels. The slicked panels exhibited more plastic shrinkage cracking than did the scratched or the back-dragged panels. My conclusion was that the scratching tool or back-dragging with the trowel opened the stucco surface up so it could breathe, and the wall dried at about the same rate. The slick surface sealed the stucco so the outer skin dried faster than the rest of the first coat, and as a result shrank and cracked.

Editor's Note: While I thought everyone knew what slicked and burned were, I was mistaken, so I am adding some definitions.

Slicked—a vernacular term referring to working stucco until fines come to the surface, resulting in a very smooth surface; this reduces the potential for mechanical bonding of subsequent coats.

Burned—a vernacular term referring to burnished stucco which occurs when a slicked stucco surface is worked once the amount of moisture at the surface of the stucco is reduced; this reduces the potential for mechanical bonding of subsequent coats.

Burnished—made shiny by rubbing or polishing.

When the stucco had cured for two days, I added the brown coat to all of the panels. After the brown coat had cured for 7 days, I broke the bond between the scratch coat and the brown coat. My findings were:

Slicked first coat—low bond,
Scratched but not cleaned—medium bond,
Scratched and cleaned—high bond,
Back-dragged—high bond.

Editor's Note: Cleaned refers to removing all lose material from the scratch coat.

Apparently when the first coat was scratched but not cleaned up before the brown coat was added; the brown coat would stick to the stucco crumbs on the wall. Since the crumbs were not bonded to the first coat very well, the bond of the second coat to the first coat was reduced. With the slicked surface, there were not as many surfaces that permitted a mechanical bond. An advantage of scratching or back-dragging is that there is not a sealed surface, so the surface of the stucco is much less likely to develop hairline cracks, which can widen and lengthen in the future. Also, there is more surface area for mechanical and chemical bonding to form.

For the stucco to gain strength, moisture needs to be present so chemical reactions can take place within the scratch coat. If the color of the scratch coat lightens within the first 48 hours, chances are the stucco will never develop the strength that could have been developed. As soon as the stucco has set so that it will not be eroded by water, misting water should be applied to the wall. Some people do this with a garden hose and that can lead to erosion problems, while misting the wall with the Hudson-type sprayer works much better. The object is to limit evaporation from the wall, rather than adding water into the wall. Most standards mention keeping the wall hydrated for 24 hours; if you're located in Houston, Texas, or New Orleans, Louisiana, that probably is just fine. If you're located in any area that is drier, then it would be better to keep the wall hydrated for 48 hours. If a brown coat is added before the 48 hours is up, the hydration of the scratch coat is accomplished by keeping the brown coat hydrated. Some people use sealers, so they don't have to moisten the wall. Some of those sealers may interfere with the formation of a bond between the scratch coat and the brown coat, or they may delay the chemical hydration of the stucco. A number of years ago I saw this on a multi-story building where an acrylic polymer was added to the stucco so that the contractor did not have to worry about moist-curing. I was called in when the stucco was three days old, and I could still mark the stucco with a thumbnail. I recommended that they quit using the admixture and started investigating what was happening. By seven days the stucco was up to a 3-day strength. The admixture salesman stated not to worry, because in another day the strength would be there. The developer was talking about tearing all of the stucco off that was slow to cure. They let me monitor the building while additional stucco was added that did not contain the admixture. Each day the admixture salesman, or his technical back-up, assured me that by tomorrow we would see the strength gain we needed. That is, they assured me until I refused to listen to them anymore. At 56 days we finally reached the strength that we should have had in 28 days. Testing in the lab determined that with a Portland cement/hydrated lime mix, the admixture did not cause any retarding effect. With several commercial masonry cements, it did not cause any retarding effect. With other masonry cements, it caused a variable retarding effect. If you are going to add anything to your stucco, test it before you start building.

It used to be that the scratch coat needed to be cured for seven days before the brown coat was applied. Then somebody came up with the idea there would not be a chemical bond between the brown coat and the scratch coat if the scratch coat had cured for seven days before the brown coat was added. We now have standards that mandate that a bonding agent be used if the scratch coat has cured for seven days or more before the brown coat is applied. I have seen no test data that clearly shows that adding a bonding agent produces a better bond than is produced by following the traditional practices that have been used for decades.

Currently *ASTM C 926* states that the brown coat can be added as soon as the scratch coat is cured sufficiently to hold the brown coat. That standard does not say who should judge whether sufficient curing has taken place. When I made a few panels and applied the brown coat as rapidly as some of the commercial plasterers were adding it, I found when taking the panels apart that there was some cracking within the scratch coat. This cracking was not evident unless the panels are taken apart, so who is making the decision that it's okay to apply the brown coat two hours and thirteen minutes after the scratch coat has been applied?

There is another reason for delaying adding the brown coat. Plastic shrinkage cracking can occur in the scratch coat within 48 hours of it being applied. Most of that cracking will occur within 24 hours. If the brown coat is applied and then the scratch coat cracks, the scratch coat will crack the brown coat. If, however, the scratch coat cracks and then the brown coat is applied, there is not much likelihood that the brown coat will crack. If the brown coat is cracked, it can be camouflaged with the finish coat, but the entire wall is not as robust as it would have been if the scratch coat had been allowed to cure for 48 hours before the brown coat was applied. Obviously, the decision is usually based on whether you want the wall to last until your check clears the bank, or whether you want the wall to last for 75 or 100 years.

We all know why the scratch coat is called the scratch coat, but why is the brown coat called a brown coat? At some point in some future chapter I will furnish the answer, but for now when you don't have anything better to do, try to figure it out.

After the brown coat has been applied, it needs to be moist-cured just like the scratch coat. Most standards state that the brown coat needs to be cured for seven days before the finish coat is applied. The reason for this is that if efflorescence is going to be surfacing, much of it will have surfaced within seven days. Then the efflorescence is camouflaged by the finish coat. Yes, there will be efflorescence surfacing from the finish coat, but the amount will be considerably less than coming from the combined scratch coat and brown coat.

Finishing can be anything from the traditional 1/8″of hard-coat stucco, to an acrylic-sanded mix, to an elastomeric finish, to a watered-down latex paint. For the final finish, you normally get what you pay for, and a watered-down latex paint, although very common in Florida a few years ago, is not really what you want. More on this subject will be discussed in the chapter on finishing.

Chapter 4
One-Coat Stucco

Most people who deal with stucco have heard of one-coat stucco. I'm here to tell you that one-coat stucco does not exist. It is a misnomer. Nevertheless, it is so ingrained in our construction industry that no matter how much I scream and holler things will not change. So I will stop protesting and use the word like everybody else does. One-coat stucco basically combines the scratch coat and the brown coat into a single 3/8″ thick coat. It still needs a finish, so it is actually a two-coat system.

Building codes do not recognize one-coat stuccos. According to building code definitions, stucco is 7/8″ thick. However, building codes recognize that innovation can occur and have provided a way for innovative materials to be used, even though they are not listed in the code. The process involves developing evaluation reports for each non-code material that is to be marketed. Prior to developing an evaluation report on a category of materials (in this case one-coat stucco), the evaluation service develops an acceptance criterion. In their simplest form acceptance criterion list the tests that must be run and the results that must be achieved for a non-code-covered material to meet the intent of the building code. There are hundreds of acceptance criteria, covering hundreds of different categories of products. Even though an acceptance criteria has been adopted by the evaluation service, as things come up the acceptance criteria are modified.

Editor's Note: Criteria is often used in the industry, rather than criterion to denote one standard.

With one-coat stucco, the acceptance criteria are given in IBC *AC 10*. It lists testing that needs to be performed and criteria that must be met for a product to receive an evaluation report. For example, *AC 10* lists the following tests:

Negative wind load,
Positive wind load,
Transverse load,
Accelerated weathering, and
Freeze-thaw.

If the manufacturer wants one-hour-fire-rated walls, there are fire tests that need to be run. If a manufacturer wants an evaluation report, he submits his material to an accredited testing laboratory, and tests are performed. Then the information is submitted to the evaluation service, and the engineers of the evaluation service issue an evaluation report. While many people believe that the manufacturer writes the report, this is not the case. In one of these reports there is much boilerplate; for example, the spacing of the nails holding the lath in place is standard. The size of the nails holding the lath in place is standard. The basic differences between evaluation reports for different manufacturers' products is that some manufacturers pay for more tests to be run, especially in the fire-testing area, or the product passed more of the tests. While some people feel that the terms listed in the evaluation report are guidelines, they are not guidelines. They are mandates. If the product in question is to be considered as complying with the intent of the code, then all aspects of the evaluation report must be complied with.

There are many applicators who are very familiar with applying three-coat stucco and make the assumption that they can apply the one-coat stucco just like the three-coat stucco. This is not a good assumption. Case in point—three-coat stucco by code definition has a one-hour fire rating. In order for one-coat stucco to have a one-hour fire rating, the entire wall, including the interior sheathing, must follow the requirements of the evaluation report. Five-eighths-inch-thick interior sheetrock is required on the walls. If a builder installs 1/2″ sheetrock, because this is what he usually uses, there is no way the wall can have a one-hour fire rating unless he tears that half-inch sheetrock off and replaces it with 5/8″ sheetrock. Another requirement is that there must be fiberglass or Rockwool insulation between the studs. Using any other insulation material, no matter how cheap it is or how good it is, is not complying with the evaluation report, and thus the installation does not meet the intent of the building code.

Because many applicators were not following the terms listed in the evaluation reports, the evaluation services mandated that the manufacturers approve the applicators. Some manufacturers approve everybody who applies for approval. Others require the applicator to take a class, or sign a document that they have studied the evaluation report and will comply with its terms.

Sometimes the evaluation service basically blows it. A case in point—the early one-coat evaluation reports required two-and-a-half-pound expanded metal lath or the equivalent weight of woven wire lath. However; it was pointed out to the evaluation service engineers that the total weight on the wall was much less, since we were dealing with 3/8″ to 1/2″ stucco as opposed to 7/8″of stucco, so a lighter-weight lath would be appropriate. Since stucco weighs about 1 1/2″ lb/ft^2 per 1/8″ of thickness, the weight difference was significant—6 pounds for 1/2″ thickness compared to 10 1/2 pounds for 7/8″ thickness. The evaluation service engineers bought-off on this concept, and evaluation reports were issued with 1.75-pound expanded metal lath and lighter-weight woven wire lath. More cracking occurred, so the evaluation service, after studying the issue, mandated that 2.5-pound expanded metal lath and heavier woven wire lath be used. All evaluation reports were amended, and the amount of cracking was reduced. However, there are still plasterers who have not read the evaluation reports and are continuing to use 1.75-pound metal lath.

Are one-coat stuccos as good as three-coat stuccos? This is a question that has been asked by numerous people, but in order to get a reasonable answer, a number of qualifiers need to be added. If a one-coat stucco is applied according to the terms of the evaluation report, by definition the one-coat stucco will be equal to the three-coat stucco that was envisioned in the acceptance criteria. Does this always happen? No. Is there a reasonable explanation as to why it does not happen? No. Are there three-coat stuccos that are superior to other three-coat stuccos? Yes. Do some one-coat stucco manufacturers produce a better product than the other one-coat stucco manufacturers? Yes. Do some one-coat manufacturers provide better technical support than others? Yes. Are there some areas where one-coat stuccos are not appropriate? Yes. Are there one-coat applicators who have never read an evaluation report concerning one-coat stucco? Yes. Are there applicators, contractors, builders, and architects who believe that all one-coat stuccos are the same? Yes. Are there some applicators, contractors, builders, and architects who have no clue as to what they are doing concerning one-coat stucco? Yes.

One-coat stuccos are a wonderful alternative to three-coat stuccos if they are used by people who understand them and who will not take shortcuts. One-coat stucco is not robust enough to handle unauthorized shortcuts.

A problem I see with the one-coat evaluation reports is that they do not mandate that the one-coat stucco be finish-coated with anything to protect it. While I understand why the evaluation service does this, it leaves a loop-hole that many applicators and builders love to crawl through. They feel that if the finish coat is not mandated by the evaluation report, they can use diluted latex paint as a finish coat.

Traditionally, the manufacturers of stucco did not furnish warranties. The applicators warrantied their work. Then when one-coat stucco came on the market, some people were leery about using a new product, so one of the early one-coat manufacturers added a ten-year warranty to his product in order to buy his way into the Florida market. Before long, every one-coat manufacturer who wanted to enter the Florida one-coat market needed to supply a warranty. As a result, there developed a greater demand for warranties for one-coat work in states close to Florida than in states further away from Florida. As the market has matured, the concept of requesting warranties in Florida has increased. More and more builders there now ask for warranties for three-coat stucco work.

Different manufacturers use warranties in different manners. The producers of acrylic finish coatings often use the warranties to sell their finish coats, and to ensure that their one-coat products are finish coated.

Example:

If you use our one-coat stucco, and you apply our finish coat, you can receive a 5-year warranty.

If you use our one-coat stucco, and you apply our finish coat and our sealer, you can receive a ten-year warranty.

Those manufacturers who do not have a line of finish coats have more trouble getting adequate protection of the one-coat stucco. The Chapter on Warranties covers warranties in greater depth.

Back in 2003 I wrote an article entitled *What Well-Meaning Professionals Do To Degrade One-Coat Stucco*. I used it as a handout during AIA continuing education classes that I taught. With some modifications it was split into two parts and published by ***Walls and Ceilings*** in November 2005 and in February 2006. Over the years I received over 400 comments from people who told me that I did not know what I was talking about. I also received over 400 comments from people who stated that it was about time somebody published something like this. If a writer can please as many people as he irritates, I suspect that the writing is causing people to think. Getting people to think is the whole idea behind writing. The article will be part of ***Forensics of Stucco*** when that part of this series is published. In the meantime you can find it online by Googling: "Degradation of One-Coat Stucco." Last time I checked, the article was the first two hits. Stop everything, and read the article.

Now that you have read the article, are you irritated that I wrote the article, or are you irritated because I delayed so long in writing the article?

Before closing this chapter, I have one story to relate. On numerous occasions I received questions about how to produce one-hour firewalls after 1/2″-thick inside sheathing was installed. On one occasion a contractor who had installed such a wall had the specifications pointed out to him. The specifications mandated a three-hour firewall. He wanted me to furnish him with a method, at minimal extra cost, to convert a bastardized one-hour firewall into a three-hour firewall. I was not able to help him.

Chapter 5
EIFS & Acrylic Stucco

After the end of World War II, Europe and especially Germany had many buildings that were seriously damaged and needed to be rehabilitated. A process was developed which we now refer to as EIFS. That stands for external or exterior insulation and finish system. The process was extremely successful, and tens of thousands of buildings were rehabilitated so that they ended up with an excellent-looking stucco-type façade. In Germany and in the rest of Europe there were few, if any, failures. When the process was brought to the United States, it worked at first, and then problems started developing.

The external insulation and finish system consists of a layer of Styrofoam fastened to the building, a coating of latex-modified cement surrounding a fiberglass cloth, and a finish coat that is often referred to as a texture coat. Sometimes this Styrofoam was attached to the building frame or other substrate with an adhesive. At other times fasteners, such as screws or nails, were used. At still other times, both adhesive and fasteners were used. Traditionally, the latex-modified cement was produced on the job site by mixing one 5-gallon bucket of a latex base with one bag of Portland cement.

Latex base is any polymer which, when mixed with Portland cement, modifies the properties of the Portland cement paste in a positive manner. Back in the 1980s I used a lot of poly vinyl alcohol and even used the cheapest latex paint I could find. Currently, many of the latex bases used are acrylic polymers.

This mixture would be troweled onto the Styrofoam. While it was still wet, fiberglass mesh would be added to the wall. Then a second later of the latex-modified cement would be troweled onto the wall. After the latex-modified cement had cured, a finish coat was added. That finish coat might be latex paint or a texture coating. The simplest textured coatings are latex paint with sand added to them. Originally, several different latex polymers were utilized, but the industry fairly quickly standardized on acrylic polymers. Different textures were produced using different sizes of sand. This produced a beautiful stucco-like surface. When it was introduced into the United States, it immediately became popular.

One of the reasons that the system became popular, besides looking sharp, was its energy-conserving characteristics. One inch to 2″ of Styrofoam was placed on the exterior of the building. A 1/8″ layer weighing about 1 1/2 lbs. /ft^2 of latex cement material was used to cover it. When the sun heated that skin, it would heat up quite rapidly, but would not hold much heat because it had little thermal mass. Since heat could not readily pass through the Styrofoam layer, as heat would build up on the surface, more and more of it would be reflected back into the air. A brick wall has a much greater mass. As a result, more energy from the sun can be absorbed into the brick wall during the day without causing the temperature of the brick to rise more than a few degrees Fahrenheit. Then at night that heat would transfer into the inside of the house. This resulted in a lower heat gain to the EIFS structure than to the brick structure. Where summer cooling is a primary factor, the EIFS structure would be cheaper to cool.

One of the first problems that developed with EIFS concerned the sand that was used in the finish coat. More than one manufacturer formulated its product with washed sand. After all, washed sand would do the job, and it was cheaper than marble sand and other specialty sands that were available. Some, but not all, of that cheaper washed sand contained bits of pyrite. Pyrite is an iron sulfide, specifically ferrous sulfide, with a formula of FeS_2. This is the sparkly crystal that is often referred to as fool's gold. Since the pyrites usually make up a small percentage of the sand particles, they are usually not noticed. But after it is in the wall, the pyrite begins to rust (oxidize).

Pyrite oxidizes from ferrous sulfide (FeS_2) to ferric sulfide (FeS). This is a problem, because ferric sulfide is black in color, and is very slightly soluble. After a few months, black spots develop on the wall. Then as the ferric sulfide is slowly dissolved and runs down the wall, a black streak develops below the spot, imparting a permanent stain.

For some reason, building owners do not appreciate this change. About the only way to take care of this change was to go in, dig the pyrite particles out, and refinish the wall. A cottage industry developed for this purpose. Hundreds, and possibly thousands, of people made a reasonably good living digging for pyrite. Some of those entrepreneurs had a sense of humor. They knew that the pyrite was also called fool's gold, so they referred to themselves in private as "gold miners." Meanwhile finish coat manufacturers were scurrying around trying to find sources of sand that were free of pyrite. One of the sources on the East Coast was Georgia Marble Company. They developed sand grades that were specifically designed for the different texture requirements of the EIFS coating. By furnishing certified pyrite-free sand with specific grain sizes, and working with the coating manufacturers, standard textures became available and the black streaks disappeared. Georgia Marble Company became the primary source of EIFS texture sand in the eastern half of the United States. Suddenly the pyrite-digging cottage industry disappeared. Other problems developed however.

While the pyrite problem was found in both residential and commercial structures which had been coated with a finish-coat that contained pyrite, the next problem that developed was much more prevalent in residential construction than in commercial construction. To give a better understanding of the problem, we need to speak briefly of barrier walls and drainage walls. If you were to look at a brick veneer wall attached to 2×4 studs, you would find the mortar is periodically omitted in the vertical (head) joints between the brick in the bottom course. These are open head joints are called weep holes and are there to provide a path for water that gets behind the brick to exit the wall. Although a brick wall is not purposely designed to allow water to pass through it from the outside, this does happen. Because it can happen, the brick wall is designed so that water that does get through passes to the outside through the weep holes at the bottom of the wall. This type of wall is called a drainage wall, because it keeps water out using drainage details.

Conventional stucco has a WRB behind the lath. The stucco run ends with a casing bead or weep screed that allows the water that gets through the stucco to drain out. This is also a drainage wall. Occasionally one sees stucco carried down onto the foundation and attached directly to the foundation, without allowing for the drainage. While I have not found any standards that allow this, it happens periodically. This changes the wall into a so-called "barrier wall," because it now must keep water out using its thickness, rather than using drainage details. EIFS was used originally used as a barrier wall. With the conventional stucco, which can breathe, water that got behind the stucco could get to the outside. Styrofoam does not breathe very well, neither does latex-modified cement, and neither did the original latex-based texture coats. So if water got behind the façade, it could not go out as vapor and it could not drain out the bottom of the wall. All it could do was sit there or move to the inside of the building. At this time vinyl wallpaper was fairly common in kitchens and bathrooms. Vinyl wallpaper was not needed for a problem to develop, but it sure could enhance the problem since it does not transmit water vapor. If water got within a wall and could not get out, the 2 × 4 studs in the wall would absorb the water. As they absorb water they change slightly in size. This was not much of a problem. But when the moisture content increased passed about 70%, mold and fungus spores started growing. Camouflaged behind vinyl wallpaper 2 × 4 studs could rot without any outward sign until the damage had been done. In other parts of the structure, where vinyl wallpaper was not being used but where good air circulation was blocked, such as behind sofas, mold could develop on the paper of the gypsum wallboard.

Numerous different mold and fungi could grow. The mold often seen on the insides of houses was usually a fruiting body that produced mold spores, whose purpose in life was to spread their species. Some people have problems with some mold spores. I will not get into the argument as to whether or not these spores cause allergic reactions. Nor will I get into the argument as to whether or not black mold is toxic. Every time I have mentioned allergic reactions and toxicity, I find that there are a plethora of definitions of these terms, and otherwise-rational people have a tendency to scream and holler if their definition is not the one that is used.

Let me explain further. Many people end up with severe diarrhea when they drink milk. This is often referred to by the sufferers as an allergic reaction, but medical personnel state, however, it is not an allergic reaction because it does not cause anaphylactic shock. Rather it is simply intolerance to lactose, caused by a missing enzyme. Try explaining that to a preteen who is bloated and can't get off the pot because of severe diarrhea. The same problem of definitions occurs when we talk people's reactions to molds. What I will state is that some mold spores cause some people severe problems.

As mold grows in the 2 × 4 studs, the studs become weaker. The process is often referred to as dry rot. If the moisture content in the studs were to drop, the growth of the mold would stop. After the mold has become established in the wood, it can start growing again at about 50% moisture content, rather than 70%, that is needed to start growing in the beginning. If there is plywood sheathing, or OSB sheathing, the mold will grow in it. As mentioned above, the sheetrock or wallboard on the inside of the structure has paper on both sides of it. This paper provides a food source for the mold to grow. Each 2 × 4 stud is sitting on a horizontal 2 × 4 that is referred to as the bottom plate. The bottom plate is normally bolted to the slab or securely nailed to wood framing. When the bottom plate gets weak from mold growth, there literally is nothing except the weight of the structure holding the house on the foundation.

In order to correct the problem, the walls need to be torn open, damaged 2 × 4 studs and the bottom plate need to be replaced, and the EIFS needs to be removed and replaced.

Oftentimes the internal sheathing needs to be replaced, also. But this chapter is not about the problems of remediation, this chapter is about this façade and problems that can happen.

We have mentioned water getting into the wall, but we have not discussed the different ways that water can get in. First off, there are penetrations in the wall—things like windows that are not adequately sealed. Water works its way around a window frame during a rainstorm, and it has no way of getting out. Some window frames have a screw in the top, or two screws in the top, and if they are not caulked, and the window header is not adequately flashed, water can enter the window frame, run down the inside of the window frame, exit and enter into the wall. If, during the summer, humidity is high and the air conditioning is running inside the house, water vapor can work its way behind the EIFS coating and condense against the cool sheetrock in the wall cavity. The water has no way to get out. Sometimes there are roof leaks where water works its way into the wall cavity. Again the water has no way to get out. Occasionally, plumbing leaks occur; usually the wall has to be opened up to correct them so that water has a way to get out. So you see, when you have a barrier system, a way for water to get into the system, but no way for the water to get out, problems will occur.

The external insulation and finish system has traditionally worked much better in commercial construction than it has in residential construction. The main reason is that in commercial construction there are more inspectors, and waterproofing of the windows is handled by a waterproofing contractor, rather than a laborer with a 79 cent tube of caulk and no idea what he is doing. All walls move. The movement is most pronounced between dissimilar materials, such as a window and the façade around the window. When the movement occurs between the window and the façade around it, any simple caulk joint at the interface has a tendency to crack or will de-bond from either the window or the coating. Since it's seldom, if ever, that a homeowner inspects and replaces caulk, water starts getting into the building, and it cannot get out.

Remember, if you take shortcuts with EIFS, a problem will probably develop; as with one-coat stucco, shortcuts are not authorized.

I find it fascinating that the same EIFS product that is not fit to use on homes, is fit to use on commercial structures. Could it possibly be the applicators who are at fault? Could it possibly be the homeowners who do not maintain their buildings who are at fault?

After the mold problems with EIFS, drainage wall systems have been designed. This has limited a lot of the potential for mold growth within the wall, because it allows water to exit. However, many building authorities and many insurance companies are against the use of this system in residential construction.

The system that we have been discussing is sometimes called synthetic stucco. This is a news-media term for it. It is not stucco, but like the one-coat stucco, I will never be able to convince the news media that it is an external insulation and finish system and that name does not include the word stucco. The finish coating that goes onto the system is normally an acrylic texture coat. It is sometimes called acrylic stucco and at other times called synthetic stucco. As mentioned before, the simplest explanation of this product is that it is an excellent grade of acrylic paint with texture sand blended with it. It is usually sold in 5-gallon plastic containers. Each container can be tinted just like paint can be tinted. No matter how careful the tinting process is, there can be slight variations from bucket to bucket, just like there can be slight variations between cans of paint that have been tinted. Each manufacturer furnishes instructions on blending buckets of their material together, so that the entire wall is uniform. Through the remainder of this document I will refer to this product as acrylic stucco. It serves the same purpose that colored cement stucco finish coats serve, but has several characteristics which they do not have. If you want the same color for the entire wall, it is easier to obtain it with an acrylic stucco. The acrylic stuccos cover cracks better than the cement stuccos. The acrylic stuccos are less likely to chip than the cement stuccos. If you want what some architects describe as a mottled old-world look, you don't want to use the acrylic stucco.

There is one other type of acrylic stucco—the elastomeric stucco. This material is made using an elastomeric acrylic base, and like the acrylic stuccos discussed above, is basically an excellent paint with texture sand mixed with it. The elastomeric base allows it to stretch, like a rubber band. This helps bridge cracks that move slightly. The elastomeric polymer tends to be thicker than the regular acrylic polymer, so sand particles carry a little more of the paint. This tends to mute the texture to some extent; however, if there are cracks in the substrate, a little muting of texture is not a big price to pay.

Chapter 6
Doing It Yourself

Stuccoing is not easy, and amateurs do work that looks less than professional. Sometimes the best and least-expensive decision is to hire the job done. At least, get an experienced person in to help and to show you how to do the work. I spray base coats, but I'm an amateur when it comes to applying most of the finish coats. The hardest finish is a perfectly-flat finish. Every imperfection shows up.

Materials and Supplies

That having been said, in this section we will talk about how you can perform an acceptable job of stuccoing. The first thing you should do is to read the rest of the chapters in this section. Then you need to assemble tools and supplies. Here are lists of what you will need:

Practice Panels

Two, and preferably 4, pieces of 1/2″ plywood or 1/2″ OSB (oriented-strand board) that are 2 ft wide and 4 ft high,

WRB or polyethylene plastic,

Roofing nails at least five-eighths inches long (but inch-and-a-quarter nails are much easier on your fingers),

12 feet of 3/4″ casing bead (see chapter on Lath Accessories for an illustration) per panel to be built,

4 feet of 3/4″ control joint (see chapter on Lath Accessories for an illustration),

2.5-lb/yd², self-furred, galvanized, expanded metal lath,
Hammer,
Tin snips, or lath shears,
Circular saw with a plywood blade and with a metal cutting blade,
Tie wire, and
Pliers with wire cutters.

Stucco Tools

A mortar tub or a wheelbarrow,

A stucco-type trowel (and for this exercise I would recommend a 12″ × 4″ pool trowel—that is a trowel where the ends are rounded),

A 12″ × 12″ hawk (they make larger ones, but a small hawk is easier to learn to use),

A concrete hoe (if it does not have a large blade and holes in the blade, it will not be efficient for mixing stucco),

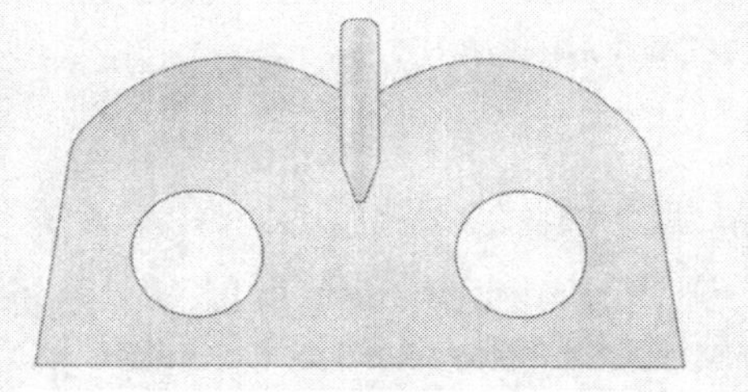

Figure 1. Blade of concrete hoe

At least two 5-gallon buckets,

A number 2 square shovel, and

One darby—this is a screed board and may be as simple as a 1×4 about 3 ft long.

Stucco Mix

One bag of Portland cement,
One bag Type S hydrated lime,
4 ft³ of loose, damp sand,
Water.

Personal Protection

Safety glasses or goggles—**wear them**.

Gloves—Stucco is alkaline and will burn or dry out your hands. Start off using substantial waterproof gloves; and if you decide to go with less protection at a later date, you can.

Apron—When you start stuccoing, you're going to be messy. Getting stucco on your pant legs and having the alkali from the stucco held there by the material in your pants can result in chemical burns to your legs. Start off with an apron, even if it's simply a piece of black plastic tied around your waist with a rope.

Practice Panels

Now that you have all of your materials and supplies, start off by building your practice panels. I would recommend that they be 2-feet wide and 4-feet tall. You can build them out of half-inch plywood or OSB. You can use thinner plywood or OSB and brace the panels with 2 × 4s. These do not need to be permanent panels; and if you want, you can salvage the lumber that you used to make them. The following assumes that you will be using 1/2″ plywood or OSB. Normally you would apply a WRB. In this case, if you have either clear or black polyethylene plastic available, you can use that. On any structure you build, do not use polyethylene plastic. In ***The Stucco Book—Forensics & Repair*** at least one nightmare resulting from the use of polyethylene plastic will be discussed. Apply the WRB or polyethylene plastic to the boards. Use a minimum number of roofing nails to hold it in place. Apply 3/4″ casing bead around the edge of each panel.

Editor's Note: Polyethylene plastic is allowed on the practice panels so the plywood base can be salvaged.

While applying the casing bead, you might as well learn how to do a mitered corner. There are two ways. One is to cut the casing bead at 45°, attach it to the next piece of casing bead at 45°, and attach. A much-neater corner can be made by not cutting the casing bead into two pieces but by taking a notch out from the webbing side of the casing bead of a little bit more than

90°. In order to get the casing bead to bend without deforming the outer edge of the casing bead, the screed point also needs to be cut. When the casing bead is attached and bent, and the second leg is attached, you end up with a much-neater mitered joint than if you had cut completely through the casing bead. Try both ways—decide which way you're going to use when you're building a structure. When you attach the casing bead, use a minimum number of roofing nails to hold it in place. With one of the panels, attach a control joint down the center. Attach the control joint by using tie-wire to tie it top and bottom to the casing bead. The control joint comes with a film of plastic over the top of the "M" shape in the control joint. Remove the plastic from half of the control joint, and leave it in place for the other half.

Now cut and attach the lath. Most of the pieces of lath need to be just under 2′ long, so they can easily fit between the vertical casing beads. Lath comes in sheets that are 8′ long, and 27″ high. Usually the term "wide" is used rather than "high," but I'm using "high" to ensure that you place the lath in the proper direction of the panels. If you cut pieces so they are 23″ × 27″ and apply them to an area that is 23 1/2″ × 47 1/2″, you will have an overlap of about seven inches. For at least one of the panels leave that overlap, but with one panel go ahead and trim it so you have one inch of overlap. For the panel that has the control joint, the lath will need to be about 11″ long by 27″ high. With the panels without the control joint, wire tie the two pieces of lath together at the horizontal joint at the center of the panel.

You may cut the lath with tin snips, a pair of lath shearers, or, if you will promise me you will use safety glasses, you may cut the lath with a circular saw and a metal cutting blade. If you are going to do the latter, set up a working table so that the lath can lie on the table and across two 2 × 4s with a space between them. The space shows where the lath needs to be cut and ensures that you're not going to be running the saw blade in the dirt or losing track of where you should be cutting. Set the blade so it will penetrate the lath about 1/4″. I would recommend that you cut only one sheet of lath at a time. With experience, a number of sheets can be cut at the same time.

It is now time to determine which side is front and which side is back. It is also necessary to decide which way is right-side-up, and which way is upside-down. If you have self-furred lath, and you should have self-furred lath, you will notice periodic dimples on the lath as if a ball peen hammer had hit the lath. These protrusions go to the back. They are used to hold the lath away from the wall. With the small panel you are using for these practice panels, you may end up with a few pieces that don't have any dimples. When you hang the lath on the wall, it should be such that the expanded metal lath forms little baskets. When you run your hand **down** the lath, the lath feels smooth; when you run your hand **up** the lath, it feels rougher. If you hang the lath upside-down, there will be more of a tendency for the stucco to fall off.

The lath should be nailed with roofing nails every 7″ along both vertical casing beads. With the horizontal top and bottom casing beads, apply one roofing nail in the center of each. Now comes the fun part—you get to wire-tie the control joint to the lath. You will place a wire tie every seven inches along both sides of the control joint. To do this, take a pair of pliers (I like linemen's pliers) and then bend approximately the last half-inch of tie wire into a J. Insert the wire through the lath and the casing bead, and pull it back out, so it catches on both casing bead and lath. With the pliers, twist the wire to make a connection, cut off the end of the wire, and bend another J. Move over 7″ and repeat the process.

While you can set a practice panel up against a wall so that the base is 6″ from the wall and use that as your stuccoing surface, if you have a few years on you, your knees will thank you if you attach the panel with the base 2′ to 3′ off the ground. However the panel is sitting, it should be stable and not move.

Mixing Stucco

It is now time to mix up some stucco and play with it. Stucco is traditionally mixed by volume, so we will keep up with tradition. I would recommend that, if at all possible, you work with a mixture that contains 1 part Portland cement, ½ part Type S hydrated lime, and 4 1/2 parts of loose, damp stucco sand. If you cannot get stucco sand, use masonry sand. If you decide you want to use a pre-sanded stucco mix or a pre-sanded mortar mix, go ahead and do so. Your results will probably not be as good as you would have with the mix I have recommended.

With the first batch of stucco you make, you are simply going to learn how to mix stucco in a wheelbarrow, and then dab a little on the practice panels. Hopefully you have a 6 ft^3 contractor-type wheelbarrow. This wheelbarrow is so much easier to work with than the garden-type wheelbarrows that are periodically used. It is also easier to use than a mortar box or mortar tub. Do not think that such a wheelbarrow will hold 6 ft^3 of material. Actually it will, until you try to move it. If you're not very adept at using wheelbarrows, you will probably turn it over. For this first batch use one shovel of Portland cement, 1/2 shovel of hydrated lime, and 4 1/2″ shovels of loose, damp stucco sand. Always fill the shovel to the same level. If you're not comfortable measuring shovels, get a 3-pound coffee can and use that. For measuring both your Portland cement and the Type S hydrated lime, you need to tap the side of the can to compact the powder. Start off by placing the sand in the wheelbarrow. Spread it out over the bottom. Then add the Portland cement. Spread it out over the sand. Then add the Type S hydrated lime. Spread it out over the Portland cement. Using the concrete hoe, and standing in front of the wheelbarrow, take 3/8″ cuts into the three layers of material and pull each cut towards you. This'll take a little while, but if the materials are not evenly mixed when they are dry, it will be extremely difficult to get them evenly mixed when they are wet. If your stucco is not evenly mixed, there will be a problem applying it to the wall.

After you get all of the materials in your wheelbarrow moved to the front end of the wheelbarrow, move to the back end and repeat the process. By this time the mixture in the wheelbarrow will visually appear to be well-mixed. Move to the front of the wheelbarrow and repeat the process. Then it is time to add water. Ideally, you are going to want to add the amount of water in weight that is equal to about 50% of the combined weight of the Portland cement and the hydrated lime. Since you will not have a scale to weigh the components, you're going to learn to estimate the amount of water needed. Fill a 5-gallon bucket with water. Slop some of the water into the back of the wheelbarrow. Standing at the back of the wheelbarrow, using the same 3/8" chops, mix enough of the dry stucco mixture with the water to make a paste. Slop some more water into the area between the paste and the dry stucco mix. Continue to do this until all of the stucco mix is in paste form. The ideal consistency is about like toothpaste. If you think you are close to correct, move to the other end of the wheelbarrow, using the same chopping motion, and mix the paste.

Now is the time to determine whether you have the appropriate amount of water in the mix. Utilizing the back side of the trowel (the side with the handle), pick up some of the stucco mix and place it on the hawk. Holding the hawk at a 45° angle and placing the trowel at the bottom edge of the hawk, also at a 45° angle, sweep the trowel across the face of the hawk to load the stucco onto the face of the trowel. During the process, move the hawk so it becomes more vertical and the face of the trowel becomes more horizontal. You now have a loaded trowel. Not loaded with much, but loaded. Throw the stucco back in the wheelbarrow and do it again, and do it again, and do it again. Now load the hawk with twice as much stucco, and repeat the process. Now load the trowel and then start tipping it. The stucco should stay on the trowel when the trowel has been tipped to about 30° from horizontal, but it should slide off when it is 45° from horizontal. While it is on the trowel, it should hold its shape, and water should not leak from the pile of stucco. If the stucco on your trowel as it is held horizontally does not hold its shape or if the water leaks from the pile of stucco, you have too much water in the mix. Dump the mix where it will not cause a problem, wash the wheelbarrow and other tools, and start over.

Editor's Note: If you were on a job, you would probably add more dry material, but with this exercise you are learning to add the correct amount of water.

If the stucco sticks on the trowel when the trowel is held at a 45° angle, it probably does not have quite as much water in it as it should have. If that is the case, add just a little bit more water, probably about a cup with this size batch, and using the chopping method, mix the stucco. Run your test again. When you are satisfied that the stucco is well-mixed and has the right consistency, measure how much water was used from the 5-gallon bucket using a tape measure. Write this number down. If you mix a batch that contains three shovels of Portland cement, one and a half-shovels of Type S hydrated lime, and 13 1/2 shovels of loose, damp stucco sand, you will need to use three times this amount of water. When you're mixing such a batch, don't put all of the water in at first, because the sand might contain a different amount of moisture. Use about 80% of that amount of the water, mix the stucco, and add water as needed. When the mix is correct, it is called "mud."

Applying Stucco

Load your hawk with enough mud to fill your trowel. Approach your practice panel, fill your trowel from the hawk, place the trowel at about a 45° angle from horizontal, against the lower portion of the practice panel, and move the trowel upward. As you move the trowel upward, drag the trailing edge of the trowel against the lath, and gradually rotate the face of the trowel until it is about 10° from the practice panel. As the mud on the trowel is used, gradually increase the angle of the trowel until it is about 20° from the practice panel. As you complete your stroke, move the trowel in a sweeping motion to one side (normally to the left if you're right-handed). At this point that little extra movement will not mean anything or help you; but if you develop it now, when you are adding brown coat and finish coat, it will help break the suction between the trowel and the mud so you do not pull that stucco off the wall. Repeat the process.

With each trowel-full of mud, overlap the previous troweled-on material until you have the entire surface coated. When you start on your second panel, try to hold your trowel out from the expanded metal lath about 1/8″. This will give you 3/8″ of stucco on the panel. When you move to the panel with the control joint in the middle, you will find why every plasterer has a few short trowels. You don't get to use a short trowel; you get to use the end of the trowel to apply the mud.

When you get a panel coated with the first coat, after it has dried for 30 minutes to an hour, use a nail and place horizontal scratches into the surface. This is to show you that if you want to apply very much stucco, a scratching tool makes a lot of sense.

Normally we would allow the stucco to moist-cure for two days before adding the final coat. On this one occasion it is okay after you have completed your three or four panels to come back and start applying the brown coat.

Fill your hawk. Standing over the wheelbarrow, fill your trowel and dump the trowel of mud into the wheelbarrow. Again fill your trowel, and again dump the trowel of mud into the wheelbarrow. In the process, you may end up dumping mud off of the hawk, or you may find that you do not fill your trowel. Keep practicing. When you think you have the process down, apply a hawk-full of mud to your practice panel. As you are doing it, you will probably end up spilling some mud. Holding your hawk horizontal while applying mud is a skill that takes practice before you master it.

Apply the brown coat in a similar manner to the scratch coat, and bring this total stucco coat out to where it is even with the outer edge of the casing bead. After getting a panel half-finished, use the darby to level out the stucco by starting at the bottom and working the darby back and forth as it is pressed against the outer edge of the casing bead, and as you move it up the wall. Apply all of the stucco that you mixed to the practice panels.

The Learning Curve

Let the stucco cure for at least 3 days. Then go to the panel that has the control joint. Peel off the tape, and clean out the groove in the casing bead. By the time you get it cleaned out, you will have learned that you never, ever under any circumstances remove that tape prior to applying stucco. There are people who fill that groove in a control joint with caulk—this defeats the reason for having a control joint.

You now have the knowledge to apply stucco as a do-it-yourselfer. In fact, you have more knowledge than some professionals.

Chapter 7
Framing & Sheathing

Most people have a problem plastering on air. Normally we need something to support the stucco, since stucco in and of itself is usually not considered structural. Some substrates we stucco, serve not only as the building's structure, but also as the plaster base to accept the stucco. An example would be a concrete block wall. It is the structure, but it can also serve as the plaster base, so we will cover concrete block in this chapter and in the chapter on plaster bases.

Wood Frame Construction

Probably the most common method of construction in the United States is utilizing 2 × 4 wood studs. The studs for load-bearing walls are usually placed 16″ on center. Within the United States there is a large contingent of people who know how to frame a house with 2 × 4 construction. Houses go up quickly, are sheathed, and dry-in quickly.

In cooler portions of the United States, 2×6 wood studs are commonly used to allow more insulation to be placed into the walls.

A problem with wood frame construction is that not all wood studs are straight. As a result, walls are built which are not straight. When a man shows up with a trowel to plaster a wall that is not straight, he has extra work to do to straighten the wall. He does it by applying stucco thicker in some areas and thinner in others. After a good trowel man gets through, you'd never know that the frame wall was not straight.

An additional problem with wood-frame construction is that mold can start growing if the moisture level of the wood gets to about 70%. After it once starts growing in wood, the wood only needs to get up to about 50% moisture before the mold will start growing again. Those numbers of 70% and 50% can be very misleading. The wood only needs to get to those levels in a very small area. In years past, water could get in a wall and could get out of the wall. With the tools we currently have, along with human error, we often cannot keep the water from getting in the wall, but we are experts at keeping the water in the wall. It only takes someone's inattention for a few minutes while waterproofing a structure to make a mistake that will allow sufficient water to get inside of the wall to cause a problem. If the mold just grew, and used the 2×4 for support, there would not be much of a problem. However, the mold requires a food source, so it digests the sugars and cellulose that are in the wood. This causes a decrease in the ability of the wood to carry a load. In the vernacular we call this "rot."

To overcome the problems with wood frame construction, steel studs are commonly used. Steel studs are always straight (unless somebody bends them). Additionally, steel studs do not provide a food source for mold, so they do not rot. If water gets in and stays in for an extended period of time, the steel studs can end up rusting. This is not good.

Sheathing

After the studs are up, normally a sheet-type material is placed over them. This is called sheathing. When plywood is used, or oriented strand board (OSB) is used, considerable reinforcement is given to the stud wall. When one of these wood-based sheathings is used, it needs to be placed with gaps between the sheets, because wood can absorb moisture. When that happens, the wood swells. When it swells, if the sheathing is not gapped, it will buckle. Not a great big buckle, but a little buckle. If the buckle is away from the stucco, the stucco does not have the support of the sheathing behind it. If the wall is hit, a crack can result. If the buckle is towards the stucco, it can put stress on the stucco and cause a crack in the stucco. These cracks tend to be between the studs, and meander a few inches back and forth. Cracks can also occur where two panels are butted together. The expansion of the sheathing causes the edge of the sheathing to push out, resulting in the stucco cracking.

At this point it is necessary to mention a strange phenomenon that occurs on many stucco jobs. When it is mentioned to anyone at a job site that the wood-based sheathing needs to be gapped, a majority of people claim that they have never heard of this before. When it is mentioned that roof sheathing needs to be gapped, they admit having heard of that, even though they may never have applied roof sheathing. When it is pointed out that each piece of sheathing states that there needs to be at least a 1/8″ gap between the pieces of sheathing, a majority of people on the job site state that this must be something new because they have never seen such a statement on the sheathing before. Six months later going back to the same crew, there is a strong possibility that the same conversation will take place. Usually the same crew does not apply the sheathing and the plaster, but when the plaster cracks, the plasterer gets the credit for it. In many cases, the plasterer never knows that the sheathing was not gapped, unless he removes the WRB.

Until management at job sites is more concerned with quality than cheap-and-fast, we will continue to see this kind of problem. This is pointed out concerning stucco. Similar problems occur with other types of material. Management is not likely to change their priorities, unless code officials mandate a sheathing inspection.

There are other types of sheathing that do not require gapping. They include exterior type gypsum board, asphalt-impregnated fiberboard, and a number of proprietary sheathings. One of the proprietary sheathings is a gypsum board that has a fiber glassed face.

Then we move to what is often referred to as open-frame construction. EPS (expanded polystyrene) foam can be attached to the studs, as sheathing. Then the lath is attached on top of that. There are some codes that require WRB behind the EPS foam. Other codes require WRB in front of the EPS foam. There is at least one case where the EPS foam is considered a WRB if it meets certain criteria and is tongue-and-grooved. In other cases, a wood- or gypsum-based sheathing is applied, and then a layer of EPS foam is added.

Another method of open framing that I saw regularly as a kid, but don't see much of anymore, involves adding felt paper to the studs, and then adding diamond mesh lath. Stucco would then be troweled on. The felt paper would keep the stucco from passing through the lath and dropping to the ground. If the first coat was put on without a great deal of trowel pressure, after the first coat got hard, the second coat could be applied with considerably more pressure.

Concrete Block and Other Masonry

When a concrete block wall is built, especially if it is built from lightweight block, it may leak during a driving rain. One way to fix it is with block sealer. Another way to fix it is to stucco it. If the block has been sealed, there can be a problem making the stucco stick. If a person plans on stuccoing a new concrete block building, block without a sealer should be ordered and used. Some integral water-repellent admixtures cause a problem as well, so it is a good idea to find out before ordering the block.

Sometimes an old concrete block wall is rehabilitated by stuccoing. As long as it doesn't have paint on it, or a sealer, and is in good condition physically, it makes a wonderful frame to hold stucco.

If a WRB is needed, then lath is needed. If a WRB is not needed and stucco will bond to the block, a better job can be achieved with direct application of the mud to the block.

A nice thing about concrete block and brick construction that does not occur with wood frame construction is that as long as the foundation is adequate, the wall does not move around very much.

Poured Concrete

When concrete is poured, there is a form release material that is added to the forms. Years ago the forms were oiled but in recent decades polymers have been developed that prevent the concrete from sticking to the forms. If some of the form release agent is left on the concrete, it can interfere with the stucco bonding to the concrete. A lot of the poured concrete has air bubbles and other surface defects in it. Stuccoing the concrete is a way to camouflage those defects. The polymers can be removed with sand-blasting or bead-blasting. Some acids will cut some polymers. If the polymer cannot be removed, the only option is to install lath on the wall.

Editor's Note: Younger people think that recycling is something new, but we saved the oil from oil changes in vehicles to oil the forms.

Tilt-wall

Tilt-wall is another form of poured concrete, but it is poured lying down, and then it is stood up. As a result, one side of it is the bottom of the pour. Often the pour is done on a concrete slab that is covered with polyethylene plastic. This may lead to slick areas of the wall that don't bond very well. The other side of the wall is usually screeded off, and then troweled. After troweling, a curing compound is often applied. This curing compound may interfere with the bonding of the stucco.

A number of years ago, tilt-wall panels were made on-site and poured on the slab of the building where they were going to be used. In modern times, most of the tilt-wall panels are made in a factory and then trucked to the site. This has the advantage of being able to install all of the panels at one time. It has many other advantages—there is only one crane rental; quality-control procedures are much easier to implement; and there is economy of scale in pouring the tilt-wall panels.

A negative aspect of tilt-wall panels is that at regular intervals there is a vertical seam. There will be movement at that seam, so expansion joints need to be installed at each of the tilt-wall seams if the panels are very wide. Otherwise, control joints will **usually** work.

Each tilt-wall panel has several pieces of steel embedded in it. The slab has several pieces of steel embedded in it. When the panels are stood up with a crane, the pieces of steel are aligned, so they can be welded together, and welded to the slab. Don't rush the job. If the tilt-wall panels have not cured as much as the slab, they will continue to shrink and the shrinkage will pop the welds.

Chapter 8
Vapor & Weather Barriers

Over the last 60 years I have listened to lots of comments concerning vapor barriers and weather-resistive barriers (WRB). Most of those comments that I remember have been wrong. When you start looking at what vapor barriers do and what WRB do, it's no wonder that most of the world is completely lost. A WRB is designed to allow water vapor to pass through it, but to prevent liquid water from passing through it. A vapor barrier, however, is designed to prevent water vapor from passing through it. Whether you need a vapor barrier or not, and if you do, where it should be placed, is geographically determined. If you're in Chicago, there is no doubt that you should have your vapor barrier on the interior side of the walls. If you're located in New Orleans, however, you want your vapor barrier on the outside. In all cases, a WRB goes on the outside. All of this makes a lot of sense, when you start looking at what a vapor barrier does and what a WRB does. The problem is that many people confuse vapor barriers and WRB.

Vapor Barriers

The Department of Energy website has a good explanation concerning the placement of vapor barriers.

http://www.energysavers.gov/your_home/insulation_airsealing/index.cfm/mytopic=11810

The vapor barrier is placed in a wall so that moisture will not condense inside the wall and lead to problems. For example, in Chicago where you have many heating days, the vapor barrier goes on the inside. That way warm, moist air that is found inside the house during cold weather does not end up inside the walls. If it did, and cooled down, the water vapor in that wall would condense. Condensate in a wall can lead to mold growing in the wall.

We have a different situation in New Orleans, Louisiana. Here we are more concerned with summertime conditions, when humidity is high on the outside of the house, and the inside of the house is kept cool. The goal is to keep the hot, humid outside air from getting into the wall and condensing when it cools. On the inside of the house the air conditioning system keeps the humidity reduced.

While this analogy is not ideal, to picture what is happening, look at a double-paned window. Where does the condensate collect? Since it collects more on the inside during a given year in Chicago than on the outside, the vapor barrier goes on the inside. In New Orleans the condensation occurs more on the outside, so the vapor barrier goes on the outside.

For most of the US the vapor barriers go to the inside. Coastal areas (within 75 to 150 miles of the coast) from Texas to North Carolina are an exception. Additionally there is a transition zone that ranges from 50 to 75 miles wide where the vapor barrier can be omitted. If you are building in these areas, go to the Department of Energy website and learn a little more about the placement of vapor barriers. To complicate the problem, each year conditions are a little different. Then there is global warming.

Fiberglass insulation batts with paper backing usually state that the paper is a vapor barrier and should go to the inside. For most of the country this is correct. But if you're building in New Orleans, it doesn't make sense. While many building inspectors understand about the placement of vapor barriers, there are a few who are still learning. In San Antonio it is a common practice with some builders to install the paper to the inside and then use a knife to cut holes in the paper. The paper is still there to hang the insulation in the wall.

In some locations builders like to add two layers of vapor barrier. One goes to the inside and one goes to the outside. In a perfect world this would keep the inside of the wall dry. We don't have a perfect world. In some manner, and it's usually around windows and other openings, water gets between the two layers of vapor barrier. It is then trapped in there permanently. Water plus any kind of lumber equals mold growth. Several years ago I was troubleshooting a problem in the Midwest, and I found that the city's building code mandated two layers of vapor barrier. They have since changed their building code. But even if it is not in the code, on a fairly regular basis it inadvertently happens when someone puts vinyl wallpaper on the inside of a bathroom or kitchen where the vapor barrier is to the outside. If any water gets in, it has problems getting out.

Weather-Resistive Barriers

We now move to WRB. Some authorities called them water-resistive barriers. If we simply refer to them as WRB, then we don't have to decide which term is correct—water or weather. Also, if we determine that WRB can be either singular or plural, then we do not have to decide whether to add an "S" behind the letters. There are basically three classes of WRB. The first class is the felt papers. We used to have 15-pound felt and 30-pound felt. Both kinds of felt disappeared from the market, and have been replaced by No. 15 felt, and by No. 30 felt. Those two felts no longer have to weigh 15 lb/square (100 ft^2) and 30 lb/square, respectively. The second class is building paper. Grade D building paper comes in several wicking rates. 10-minute paper will prevent water from passing through for 10 minutes. 60-minute paper will prevent water from passing through for 60 minutes. These sound like awfully small numbers, but don't get too excited. Most water never gets behind the stucco. So it never touches the Grade D building paper. If any water does get behind this stucco, much of it moves down the wall and exits at the weep screed. So in many cases, 10-minute paper can adequately protect the contents of a building. That is, provided that the 10-minute paper does not have holes in it.

Those holes can be in the form of tears, nail holes, staple holes, rips from sharp edges of the lath, and inappropriately wrapping of penetrations such as windows. To help guard against these holes in the Grade D building paper or the felt paper, some jurisdictions mandate two layers of WRB. That does not eliminate water moving through nail holes and staple holes, but it substantially reduces the amount of water that passes through tears. The third class of WRB is the housewraps. I believe Tyvek was the first that became popular. Then other companies followed suit. Tyvek then came out with a product they referred to as stucco wrap. It was designed so that water could better drain between the stucco and the Tyvek. A number of people have alleged that when stucco was applied directly to the Tyvek Housewrap, pores in the housewrap became plugged, and the product became a vapor barrier, rather than a WRB. Even though I have asked to see data proving that such things happened, I have never seen any data on the subject. It could be that I was just too lazy to go hunt for data, or it could be like the Tyvek salesman told me, "That is just something our competitors say so they can get part of our business."

For sealing the house to prevent drafts, a housewrap normally does a better job than either the felt paper or the grade D building paper.

Regularly I am told that housewraps are better than felt because stucco does not stick to housewraps. I am also regularly told that felt paper is better than housewraps because stucco bonds to felt paper and the felt paper keeps it from cracking. Both are probably true; and the moon is made of green cheese, and if I can just rope it and haul it down to earth, I can feed the world for 465.34 years. Yes, stucco is less likely to crack over felt paper than over housewrap. Since there is not a drainage plane between the felt paper and the stucco, more water is held in the stucco and can migrate through the felt paper to the wooden studs.

Rainscreen

Rainscreens are systems that provide a drainage cavity behind a façade. While there are some who allege that WRB are all that are required to keep water from passing to the sheathing, in more humid areas this is not the case. This is especially true when wind accompanies rainfall events. Ten-minute Grade D building paper retards the passage of water through the paper by 10 minutes. Even if 60-minute paper is used, it only delays the passage for one hour. Double layers of Grade D paper or felt paper provide a semblance of a drainage cavity, but that system is not very efficient. Housewraps do not allow the passage of water, but housewraps are attached with nails and staples. Each place where the housewrap has been penetrated by nails, staples, or jagged sections of lath, the housewrap no longer exists. Additionally, windows and other penetrations are often not sealed well and are a source of water intrusion. If there is not an easy way for the water to exit the wall, problems can quickly develop.

Rainscreens probably are not needed in Las Vegas, but they have a definite place in the more humid portions of the planet. Already there are some builders that are mandating rainscreens. For example, Pulte Homes mandated rainscreens behind Adhered Concrete Masonry Veneer in their Southeast region.

Whether rainscreens become a mandate of local CODES will be determined by whether CODE officials' priorities are quality construction or cheap construction.

Sealing Nail & Staple Holes

Periodically, when at a construction site, I see that all nail holes and staple holes through the WRB have been caulked. This is to prevent water from passing through those penetrations. Does that actually help? Does water actually pass through those penetrations? There has been lots of conversation on the subject, but I know of only one test that was run to find out for sure.

Although I was invited to view the test, I was not there. An association was trying to develop standards for the installation of their product over a stucco scratch coat. They decided to run water penetration tests on their product and also on the scratch coat. They felt that water would not penetrate the scratch coat, and if that was the case, it was a moot issue as to whether the water would pass through their product. In the discussion it was also decided to go ahead and run a test on the wall with the WRB and the lath in place, but not the scratch coat.

The test procedure was very simple. A frame for a 4-foot-wide by 8-foot-high wall was built. OSB sheathing was attached to the frame. The sheathing had to have a vertical joint and a horizontal joint. Then the wall was covered with a housewrap. Then casing bead and lath were installed. A second wall was built and treated in the same manner and then coated with a scratch coat of stucco. A third wall was built, and after it'd been coated with a scratch coat of stucco, the additional product was attached to the wall. None of the staple holes were sealed. To run the test, a specified vacuum was applied to the backside of the wall, and a sprayer system added sufficient water to the front side so there was water cascading down the wall during the test. This simulated wind driven rain. If there were any leaks, then that vacuum would be pulling water through the wall. I believe, but do not remember, that the tests were run in triplicate.

Tests were run on the walls. The wall with the scratch coat and the wall with the scratch-coat-plus product did not leak. The wall without the scratch coat leaked. In taking the wall apart it became very obvious what had happened. Water passed through the staple holes in the housewrap. It then moved laterally to the gaps between the sheets of OSB. It then moved between the OSB and the two by fours. After it got through that final interface, it ran down the two by fours and puddled at the bottom of the wall. The engineer for the housewrap company stated he had never seen anything like that; but when he was asked, he admitted that he had never seen a test run with the lath in place, but with the wall not having been stuccoed.

What about the stucco prevented the wall from leaking? I don't know the answer, but I suspect part of it is related to the vacuum not pulling through the stucco as much as it was pulling through the staple holes. Would felt paper or grade D building paper have done just as well? Would they have done better? Without tests having been run, we can't say for sure. I suspect they wouldn't have done any better. My conclusion from the test is that if an adequate scratch coat is added to a wall, and someone can **ensure** that the scratch coat does not crack, there really isn't any need to caulk the nail holes and staple holes that go into the framing. Nail holes and staple holes that go into the field would have a much greater chance of leaking. It would appear that if a design professional is concerned about leakage through nail holes and staple holes that he/she should seriously consider requiring a rainscreen and fining the contractor for every extra nail or staple placed in the wall.

Paper-Backed Lath

Prior to the introduction of paper-backed lath, we had good craftsmen who did everything right. And there was never a problem with the stucco. If you believe that, you've probably been listening to me too much. When paper-backed lath was introduced, it changed the way we installed the lath. When using the paper-backed lath, it was no longer possible

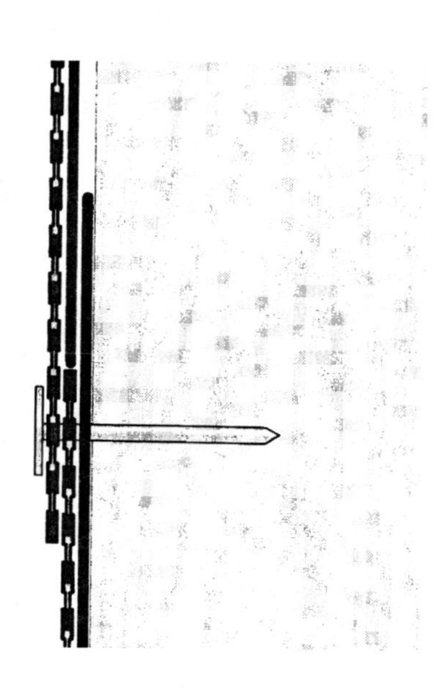

Figure 2. Lap paper to paper and lath to lath

to install the WRB, then install casing bead & control joints, and then cut the lath to length and install it. If one did that, the black paper on the back of the lath would be cut and suddenly it would not be functioning as a WRB. In just a few years, it became common to install the paper-backed lath and then to install the control joints on top of the lath without cutting the lath. This increased the number of cracks in stucco. There was an attempt to cut the lath to insert the control joints; however, it usually resulted in the paper being cut or damaged. That led to leaks in the wall. There are people who maintain the only thing paper-backed lath is good for is to be outlawed. I, however, have found a use for paper-backed lath. After a housewrap is installed and the control joints and casing bead are installed, then the paper-backed lath can be installed. The paper prevents the sharp edges of the lath from tearing the housewrap. This in my estimation is the one redeeming feature of paper-backed lath.

When installing paper-backed lath, it is necessary to lap paper to paper and metal to metal. If there is a new installer, there is a great risk that with at least some of the paper, he will install the upper sheet behind the lower sheet. If the water gets behind the stucco, then it is shortly diverted behind the WRB. This is not good.

In building free-form furniture, I have attempted to use paper-backed lath so I can shoot stucco onto the lath with the paper keeping the stucco from passing through the lath. I find that the paper is not glued on well enough to keep the stucco from going right on through. I developed a technique where I can spray lath without any paper on it, but you'll have to wait for another section of this series of books before you learn that technique.

One or Two Layers of WRB

Shortly after three hurricanes hit Orlando, Florida, in a matter of three weeks' time, I was reviewing some building codes in that area. Only one layer of WRB was required by their code.

Shortly after that I received a call from a gentleman in a suburb of Las Vegas, Nevada. He wanted to know why in his jurisdiction two layers of WRB were mandated. I guess it must be because Florida is arid and Las Vegas is very humid. Seriously, I have no idea what those people were thinking. If I were King of the World, I would mandate two layers of WRB

When You Cannot Use a WRB

Sometimes you cannot use a WRB. If you are plastering directly onto concrete block or other unit masonry, you cannot use a WRB. If the code mandates it, then you have to attach it. To do that, you must install lath. In general, a concrete or other masonry wall, or for that matter a tilt-wall that is stuccoed, is protected better, even without a WRB, than it would be if it were painted with a coat of latex paint. In these circumstances I'm not particularly concerned about the lack of a WRB.

But there are other situations where there aren't quite as many safeguards. One of these would be if one plastered on adobe block. Besides water being able to get through the stucco and into the block, which would soften as water collected in them, the adobe and the stucco have different coefficients of expansion, so over time the interface bond would probably break.

In the Southwest and in some other areas of the country, homes are being built from bales of straw. After the straw is stacked, the walls are wrapped with stucco netting. The netting is pinned to the bales, and then it is stuccoed. If there is a crack in the stucco, during a rainstorm water can flow through that crack into the straw. It may have a problem getting out. When I build with straw, I like to utilize four-foot overhangs. This does not protect the entire wall from all wind-driven rain, but it protects most of the wall from most of the wind-driven rain. I have seen others build with straw who, for curb appeal, used little or no roof overhang. If a WRB, or a vapor barrier, were to be installed on the outside of the bales, there would be no way to tie the stucco to the bales. I mentioned the use of stucco netting. When bales are stuccoed without the netting, there is a tendency for the stucco to develop cracks along the bale interfaces. The stucco netting serves to keep the water from getting to the bales. I wonder what would happen if I told a building inspector that stucco netting was my WRB?

Chapter 9
Waterproofing Penetrations

Before we can start talking about waterproofing penetrations in a wall, we need to define what a penetration is. A penetration is literally anything that interrupts the weather-resistive barrier (WRB), or as it is also called, water-resistive barrier, on a wall. This could be a window, a door, an electrical outlet, or a hose bib. So you see we have some small, seemingly-insignificant penetrations, and we have some very large ones. Roofs also have penetrations, and those penetrations may allow water to enter the top of a stucco wall. When water is around the building, it doesn't take much of a penetration before water starts entering at a rate that can cause significant damage. In constructing a wall, consider that every penetration in the wall, no matter how insignificant, is designed to flood the wall with water. If you do that, you have a mindset which can allow you to do an adequate job of waterproofing.

Rather than using the term "waterproofing," possibly we should be using the term "damp-proofing." Damp-proofing is actually more technically correct than waterproofing, because water can get through under certain circumstances. But the industry has been using the term waterproofing for so many years, that if we change the chapter title to "damp-proofing penetrations," many people would have no idea what we were talking about.

Caulks and Sealants

In this chapter, I will use the words caulk and sealant interchangeably. They are not really interchangeable, but a majority of people in the residential construction business do not recognize that there is a difference. A caulk is designed to fill a space. A sealant is designed to seal a space so that nothing will pass through. As a result sealants usually are more sticky and harder to work with. Caulks are normally loaded with mineral filler, such as calcium carbonate, and are much easier to work with and to shape so that they look good.

Many people believe that a little bead of caulk around a window is waterproofing, but we repeatedly see the beads of caulk, even caulk that's guaranteed for 50 years, fail in a few months or a few years. It is necessary in waterproofing any penetration to make the assumption that all caulk and sealants will fail. Having done forensic work on many buildings, I can assure you that that all caulk and sealants do fail. Sometimes they last six months. Sometimes they last 30 years. But they all fail. If steps are not taken back in the wall, water will get into the wall at a penetration. In the *Acknowledgements* I mentioned asking a mason how long he wanted his jobs to last; he immediately responded, "Until my check clears." Hopefully he was joking, but after examining his work, I don't think so. Looking at how penetrations are waterproofed, it appears that many people may have the same idea. Sometimes it's because they don't care, but many times it's because they don't know any better. Since we no longer have an apprentice program for training-up trades, a vast majority of tradesmen are learning on the job from people who don't know what they're doing. So much for my soap box for a little bit.

A second assumption that needs to be made is that water will travel laterally to get through an opening.

A third assumption that needs to be made is if you have windblown rain, water can travel vertically when it gets into the wall, as well as down and laterally.

Since these assumptions are normally not made in residential construction, we have to make the fourth assumption. Waterproofing as commonly practiced is designed to fail.

Designed To Fail

Anybody who has a tube caulk can caulk a window. It is simply a matter of clipping the end of the tube caulk, inserting the tube of caulk into a caulk gun, and then moving it while squeezing the trigger of the caulk gun, leaving a nice little bead of caulk. If you want to be real uptown, you can take your finger and run it along that bead of caulk and smooth it out. You not only want to caulk below the window, and at the sides of the window, you also want to caulk at the top of the window.

As soon as you have completed caulking the window, problems start developing. The window moves slightly in relationship to the stucco around it. If it is a vinyl window, the caulk does not bond extremely well to the vinyl window flanges. With the movement, hairline cracks may develop between the caulk and the window flange. Or they may develop between the stucco and the caulk. Water that gets in, cannot get out below the windows.

Caulk, as normally installed, fails. Why? Caulk has a certain amount of elasticity. Most sealants have a little bit more elasticity. Caulk fails for several reasons.

If it is bridging a gap, and the gap is a sixteenth-of-an-inch wide, and it moves so that the gap is an eighth-of-an-inch wide, and the caulk can only stretch 50%, something has to give. The caulk can separate, or it can delaminate from either the window flange or the stucco, or both.

If there is a groove between the window flange and the stucco and this groove is filled with caulk, then the caulk is attached on each side and at the back. Thicker layers of caulk are not as elastic as thin layers of caulk. The caulk attached to the back of the groove, and the caulk attached to one side of the groove, will have greater bond strength than the caulk attached to the other side of the groove.

A third reason that caulk can fail is that it doesn't get along very well with ultraviolet light. The mineral filler in the caulk is there to prevent ultraviolet light from penetrating deep into the caulk, which causes it to degrade. If the layer is very thin, the entire thickness of the caulk can be compromised. The UV light causes the polymer in the caulk to degrade. It becomes brittle, and movements of the window in relation to the stucco cause the caulk joint to fail. If caulk is thick, it will fail. If caulk is thin it will fail. Learn how to use backer rod and then use a good sealer and the job will last a lot longer before it eventually fails as all caulk and sealant joints eventually fail.

Wrapping Windows

The Traditional Method

Back in the days when 15-pound and 30-pound felt were the only WRB available, waterproofing was a little simpler. The felt paper was rolled out and attached to the wall with roofing nails. Additional pieces were attached to the wall until the felt paper reached to the top of the windows. The felt paper had nice red lines on it, so that the applicator could apply it straight, and ensure that there was an adequate lap. At windows it was cut, and about every seven to eight inches around the window frame it was nailed using roofing nails. Roofing nails have large heads and as long as hammer blows hit the head and the top of the nail is not driven through the felt paper, not much water will get through a nail hole. Since the advent of the pneumatic stapler, felt is often damaged when the staple is driven through the felt.

A strip of felt, approximately 4″ wide and at least 8″ longer than the width of the penetration, would be applied below the penetration. An easy way of developing such a strip was to take a roll of felt, mark off 4″, and saw the strip off. You never wanted to utilize your good wood saw to do this, so you kept an old saw just for this purpose. Since the asphalt in the felt would gum up the teeth of the saw, a container of gasoline and a rag were kept handy, to clean the saw teeth. Before you go report me to the Occupational Safety and Health Administration, or the safety officer on the job site, remember this was a technique that was utilized long before the Occupational Safety and Health Administration was ever thought about.

Figure 3. Flashing a window the traditional way

Then a strip was placed on each side of the window openings with the top of the strip slipping under the felt paper at the top of the opening. Then a strip was placed across the top of the opening, underneath the felt paper above the opening, but to the outside of the two strips that had been applied to the sides of the opening.

The window, with nice wide flanges, was then set into the opening and secured. At the top of the opening the flange was placed outside of the felt paper strip. Then four strips of felt paper were added around the window. Each of these strips covered the window flange. At this point the inspector would mandate that the window be removed and that it be flashed correctly. This time an additional strip of felt paper was added at the bottom of the window. A bitumen sealant was added around the window. The window was set in place and secured. Bitumen sealant was then applied to the window flanges and adjacent felt paper on the sides and top of the window. Then the two strips on the side of the window were added to the outside of the window flange. Then a strip of felt was added over the top flange.

With the window in place, a piece of metal flashing was added above the window, extending from slightly beyond the window to several inches beyond the window. The ends of the flashing were turned up about 1/4". This prevented water that hit the flashing from being diverted to the end of the flashing and going into the wall.

If the long sheet of felt paper above the window had been added with the rest of the felt paper, by this time it would have been worn out and torn. It was common practice to not add the felt paper above the window until after the windows were installed and flashed. This made adding the felt paper above the window, and to the outside of the flash, easy. The process seems very complicated, but it worked, and it worked well. After having had to tear out and start over a time or two, everyone on the job knew the correct way to flash an opening.

The same general technique was utilized in order to flash an electrical box or a hose bib.

When Housewrap Is Used

Wrapping a window where the building is being wrapped with housewrap is a little bit different. Basically, an X is cut in the window openings in the housewrap. It is wrapped inside the window frame and fastened with peel-and-stick tape is utilized.

Figure 4. Flashing a window with housewrap

Start off with a piece of tape below the sill. Then use a piece going up on either side of the window opening. A piece of peel-and-stick tape does not go above the window opening at this time. This area is referred to as over the header. Then a second layer of tape is added, moving inside of the window frame at the sill, and giving about a 2″ overlap with the piece that was below the sill. Then the same thing happens on the sides. Short pieces of tape are added to the corners to ensure that water cannot get in at the corners. Continue the process until the inside of the frame of the window is taped with overlapping pieces of tape. Add sealant, and set the window into place. Be sure to insert the top flange underneath the housewrap at the top of the window. Place a piece of tape on each side of the window, taped to both the housewrap and the window flange. Then add a piece of tape across the top of the window, tying the housewrap to the window flange.

Different Designs—Different Methods

Fifty years ago there were only a few kinds of windows. Now there are many kinds, each designed for a specific application. If you have the wrong window design for your application, waterproofing the window may be close to impossible. Recently I went on a job and observed windows designed for a vinyl siding application, but the job changed to require one-coat stucco. Five years after the windows were installed, there was considerable leakage. About the only thing that could be done was to remove the windows, tear out the one-coat stucco and the studs that had rotted, and start building back. This all happened because somebody forgot to look at the window flanges. If the manufacturer of the windows does not specifically state that the window can be used for the application you want to use them for, do not use them, even if the windows are cheap. As far as installing the windows, follow the window manufacturer's recommendation if the design professional does not provide other guidance.

Backer Rod

When installing a caulk joint, it is necessary for the lather (person whose primary job is to install lath) to be involved. He needs to leave an appropriate-sized gap between the casing bead and the window frame.

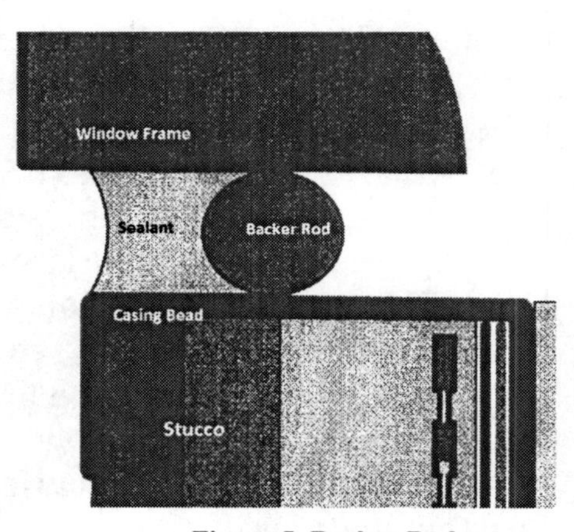

Figure 5. Backer Rod

There are some people who stucco right up to the window frame, but these are the people you really do not want on your job site, no matter how cheap they are. The casing bead is installed a measured distance from the window frame. When the wall is stucco, there will be a groove between the casing bead and the window frame. This groove is filled with a backer rod.

Backer rod comes in several different diameters, and it is a flexible plastic that is compressible. The gap should be slightly smaller than the backer rod that will be utilized. For example, if the gap is 3/8″, it would be appropriate to use 1/2″ backer rod. Once it is in the groove, it is not likely to come out unless some-one pulls it out. In order to seal the window, a bead of sealant is placed on the backer rod. It is normally tooled. This results in an hourglass-shaped bead of sealant. A wide foot of sealant is attached to the window frame, and a second wide foot of sealant is attached to the casing bead. This caulk joint can survive considerable movement of the window in relationship to the stucco. Since it's attached only on either side, it doesn't tear itself by being attached to the back. Yes, it is bonded well to the backer rod, but the backer rod is flexible and will move so stress is not placed on the bead of caulk from that attachment.

Chapter 10
Lath and Other Plaster Bases

You need a substance to attach plaster or stucco to. Such a substance is called a plaster base. When we think of plaster bases, most people think of lath. That is just the most common one. You can actually stucco on just about anything that will hold still. When I was a kid and tried to stucco my sister's cat, I had two problems—one, the cat would not hold still, and two, my sister went ballistic. In this chapter we will discuss some conventional and some unconventional plaster bases.

Burlap

During World War II, expanded metal lath was not commonly available to the civilian market. Neither was wood lath. As a result, burlap was used as a plaster base. I know several walls that were built in the Lower Rio Grande Valley of Texas using one layer of burlap plastered on both sides. In the early 1990s I was down in the Lower Rio Grande Valley with my father, and he showed me some of the walls that he built in that manner.

They had a commercial building, and they needed to divide it up into stores. This meant that they needed to install walls. They hung burlap from the ceiling, and they attached it to the floor. Then with a soupy stucco mix, and a plasterer on each side of the burlap, each started plastering. They were basically trying to get the burlap saturated with the stucco. The two plasterers had to be careful to utilize their trowels in unison. This allowed stucco to fill the mesh of the burlap. The wall was then allowed to dry for three days. The plasterers then came back and applied stucco on one side of the wall. Two days later they came back, and they added stucco to the other side of the wall. This gave them a wall about 3/4″ thick. After waiting for two more days, they applied another coat of stucco to each side of the wall and smoothed it to finish it out. Where they had wanted to put a doorway in the wall, they had framed it out with two by fours on the edge. The shorter dimension of the 2 × 4s was just a little greater than the finished wall surface, so when finishing the wall, the plasterers would bring out the plaster to the width of the door frame.

In talking to some of the people who work in the stores, they had no idea how the walls had been built, but they knew they couldn't drive a nail into them. In examining the walls, we found very few signs of cracking. If your son wants to move into his pup tent, in the backyard, on a permanent basis, maybe the two of you should consider stuccoing his pup tent.

Concrete Block

A much more common plaster base is concrete block. Concrete block go by a number of different names, including cinder block and CMU (that stands for concrete masonry unit). Concrete masonry units come in lightweight, medium-weight, and heavyweight versions.

Concrete block are stuccoed for aesthetic reasons, and to ensure that water does not pass through. The stucco can be applied directly to the block, or if a good bond is not anticipated, then a lath or wire mesh plaster base can be attached to the block. If in the factory the block were sealed, or if the block were sealed after they were in the wall, or if the block were painted, it may not be possible to develop a good bond between the stucco and the block. A quick way to determine if a sealer has been applied to the block is to spray a light coating of water on the wall. If it is absorbed immediately, it probably has not been sealed. If there is any beading of the water droplets on the wall, chances are the wall has been sealed. Sand blasting, bead blasting, and water blasting sometimes can remove the sealer. After the wall is prepared to receive the stucco, a dash bond coat of stucco applied to the wall can help the bond. Here is how to do a dash bond coat. First, wet the wall down; let it dry for 15 minutes, then with a wallpaper brush or similar instrument, splash fluid stucco onto the wall. Water in the stucco will be sucked into the wall, and in the process some of the stucco will be pulled in. Then proceed to a scratch coat. After the scratch coat has cured, a brown coat can be added.

Brick

Why would anybody want to apply stucco to a brick wall? Probably the first reason would be to change the look of the structure. A second reason would be that either the mortar between the brick or the brick themselves have shown some deterioration, or the stucco would be applied to camouflage that degradation and to prevent it from spreading.

In Mexico it is traditional to stucco brick walls. The walls were laid up to be a base for the stucco. Since most of the brick were going to be coated with stucco, it became traditional to under-burn brick. This saved a little bit on energy costs, and also provided a surface that the stucco could bond more tightly to. When handmade brick started being imported into South Texas from northern Mexico, this caused a substantial problem. Texans did not want to plaster their brick. They also loved the dusty pink color that came from under-burning the northern Mexican clays. After a year or two it was common to look at a wall and to see a half a dozen or so brick that had eroded from weathering. Mexican brick got a bad name because of poor quality. The Mexican brick manufacturers could not understand those crazy Texans who did not want to stucco their brick homes like everybody who had common sense did.

Don Halsell, who was with the Structural Clay Products Foundation of the Southwest was the star of an add campaign that the SCPF ran. He would take a bite out of an under fired

Mexican brick. In recent years we are seeing more and more homes in northern Mexico where the brick are exposed. With moving away from handmade brick to machine brick, and with more quality control, the brick coming out of Mexico are of a higher quality and no longer need to be stuccoed.

Tilt-wall

Tilt-walls can be a problem, because they consist of a high-density concrete that is trowel-finished on one side. The trowel finish leaves fewer pores for the stucco to bond to. An additional problem is that a curing compound is often added to the top side of the tilt-wall, and form-release material is often added to the bottom side. Curing compound often interferes with the bonding of stucco. Form-release materials almost always interfere with the bonding of stucco. To get a substrate suitable so stucco will bond to it, it is usually necessary to either sand blast or bead blast the surfaces. An alternative which should be considered more often than it actually is, is to talk with the tilt-wall supplier to provide surfaces that are not contaminated by a compound that will prevent the stucco from bonding.

Note the comments about de-bonding in the next section. They are applicable to de-bonding of stucco applied to tilt-wall.

Poured Concrete

Normally, both surfaces of poured concrete have form-release chemicals on them. These surfaces have not been troweled and slicked down, so there is more of a chance of having some mechanical bonding than there is with the face of a tilt-wall, but form-release chemicals must be removed before the stucco will bond to the concrete. This can be done with sand blasting. It can be done with bead blasting. Sometimes it can be done with high-pressure washing. No matter which method is utilized, the wall needs to be checked prior to adding stucco to it, to ensure that the offending chemical has been effectively removed. If it has not been effectively removed, problems can occur, not necessarily as the stucco is being applied, but 6 to 18 months later. Poured concrete is its greatest size when it is poured, or should we say when the forms are removed. From that point on, it is shrinking. Stucco has its greatest volume when it is placed on the wall. From that point on, it shrinks. Since we have shrinkage at two different rates, if there is any form-release chemical, stress can be placed on the bond. Then we have the normal diurnal and seasonal fluctuations in temperature. No matter how close the bond between two similar materials, if one is heated, the other has a delay before it heats up. In the same manner, if one is cooled, there is a delay before the other is cooled. Therefore, as we go through a typical day of sun shining on the side of the building, and then the sun going down and the environment cooling off, the stucco will change temperature faster than the concrete located behind it. Also, the stucco temperature fluctuations will be greater than the temperature fluctuations of the concrete behind it. Each time this happens, some stress is placed on the bond. That stress can ultimately lead to bond failure.

EPS and ICF

Everybody who deals with stucco knows that cement-based stucco should not be applied directly to expanded polystyrene (EPS), or to extruded polystyrene (XPS). It is okay, however, to apply latex or acrylic-modified cements to these substrates. This happens on a very regular basis. This is what EIFS is. In this section we are not discussing EIFS, but the application of conventional stucco. I have a sample panel that I stuccoed about 10 years ago. After 10 years the stucco is still firmly bonded to the EPS board. Why does the code forbid it, when I found that it worked? What I did was before I applied the stucco, I rasped the EPS. I found that sunlight deteriorates the surface of EPS; and when the surfaces are deteriorated, it is difficult to obtain a decent bond. If the surface is rasped, and the stucco is applied within 24 hours, an excellent bond can be established. If we tried to apply this to the construction industry, how many times would we have a rasped wall that was exposed to sunlight for over 24 hours, before stucco was applied to it? There just aren't enough inspectors out there to ensure that stucco is applied only to EPS that has been freshly rasped. So even though I have found it works extremely well, I'm not recommending it, except for special situations that fall outside of the building code—things like a Styrofoam doghouse or decorative pots where the core is made from Styrofoam.

About 30 years ago various insulated concrete forms came on the market. Some were made from EPS. Some were made from a mixture of cement-type materials with Styrofoam beads used as aggregate. Some were made from other types of insulation. Insulated concrete forms must be covered both inside and outside before a structure can be inhabited. It would be nice if stucco could be applied to them, but due to code restrictions, those insulated concrete forms made from EPS cannot be coated directly with stucco. It is necessary to add lath on the wall. Some of the forms have special plastic inserts to receive nails to hold the lath on the wall. Some require longer fasteners that go through the EPS and fasten the lath directly to the concrete. Both methods tend to be labor-intensive, but at the current time it's the only way to meet code.

Expanded Metal Lath

Most of the time when we speak of a plaster base, we are thinking about expanded metal lath. Expanded metal lath is covered by *ASTM Standard C 847 (Standard for Metal Lath).* This specification covers:

2.5-lb/yd^2 diamond mesh lath,
1.8-, 2.75-, and 3.4-lb/yd^2 flat ribbed lath,
3.4- and 4.0-lb/yd^2 3/8″ ribbed lath, and
4.5-lb/yd^2 sheet lath.

Lighter-weight lath, 1.75-lb/yd^2 diamond mesh lath, is utilized by tile setters. It is too light for use with stucco, even though for a time, evaluation services allowed its use as a plaster base for one-coat stuccos.

For residential construction, 2.5-lb/yd^2 expanded metal lath is most common. The other laths are utilized for ceilings, and for commercial and industrial construction.

Lath can come as galvanized lath or as black metal lath. The galvanized lath is utilized for exteriors and where moist conditions are anticipated. The black metal lath is designed to be used where dry interior conditions are anticipated. Over the years, I have seen my share of black metal lath utilized in exterior situations, because it is cheaper.

Figure 6. Self-furred lath

Lath needs to be furred out from the wall so the mud can get behind the lath and encapsulate the mesh. It used to be that this was accomplished by adding vertical 1/4″ thick pieces of wooden lath to the wall. Then when the expanded metal lath was attached to the wall, it stood out from the wall 1/4″. Then furring nails were developed, which allowed the lath to be firmly attached, but held approximately a quarter-of-an-inch from the wall. Then somebody came up with the bright idea of putting dimples on the lath. If you look at a sheet of furred lath, it appears that someone hit the lath with the ball peen hammer. When lath is attached to the wall, it stands out from the wall about 1/4″.

There is also the option of having paper-backed lath. In the section on weather-resistive barriers, I ranted against paper-backed lath. Since it is still fresh in your memory, rather than writing the rant again, I will ask that you simply remember what I said.

When expanded metal lath is hung on the wall in the correct manner, there are little pockets that appear to form. These little pockets should point in an upward direction. It helps keep the stucco on the lath. If you rub your hand across lath that is installed correctly, as you move your hand down, it feels smooth; when you move your hand up, it feels rougher.

You must endure one more story about paper-backed lath. Several of us were on the job in Houston, Texas, and the lathers needed to trim the paper on the paper-backed lath as they were installing it. Upon examining the lath, we found the paper was attached to the wrong side of the lath and positioned wrong. My companion called the lath company on his cell phone and explained the situation, and even I could hear the response. "*Expletive deleted*, I thought we caught all the truckloads of that *expletive deleted* lath."

Woven Wire Lath

Woven wire lath is often referred to as chicken wire, because superficially it looks a great deal like that chicken wire that many of the older generation are familiar with. It is covered by *ASTM C 1032 (Standard Specification for Woven Wire Plaster Base)*. It comes with openings of 1″ and 1 1/2″. Probably the most common woven wire lath is 17-gauge 1 1/2″-mesh. There are lighter weights, and at one time evaluation services allowed 21-gauge lath to be utilized for one-coat stuccos.

Woven wire lath is most commonly used in the West. The diamond mesh lath mentioned in the previous section is most commonly used in the East. Stuccoers, who are used to one, sometimes have problems switching to the other, especially if they are used to using the diamond mesh lath, and they are moving to the woven-wire lath.

Welded Wire Lath

Welded wire lath is covered by *ASTM C 933 (Standard Specification for Welded Wire Lath)*. Some of this lath comes in sheet form; others come in rolls. Some of it has formed wire and with welding can be rolled, but it does not have a memory. That makes applying it to a wall much easier. A lot of it comes self-furred.

Plastic and Fiberglass Lath

As soon as you get close to the coast, the amount of chloride (salt) in the air usually increases. If there is a breeze coming in from offshore, often it carries droplets of salt water, which are deposited on the surface of structures near the shore. This can cause deterioration in the metal lath. To counter this, companies have come up with plastic lath and fiberglass laths. *ASTM* is working on standards for these laths. Until *ASTM* completes the process, the plastic and fiberglass laths will be sold based on evaluation reports. Having tested some of these laths, I find that there is a little bit of a learning curve before one can get the most out of them. Also, it appears that in situations where there is more weight on the wall than conventional three-coat stucco, some of these laths may not be the most appropriate solution.

Having spent time on the Florida Gulf Coast, and looking at buildings that were completed fewer than six months before, and being able to see rust marks on corner beads, casing beads, weep screed, and control joints that were made from galvanized steel, there is a need for an alternative. Prior to the plastic lath and fiberglass lath, the only alternatives were zinc accessories. They were rather expensive.

Chlorides and Metal Lath

In the previous section, mention was made of salt-water-induced deterioration of lath accessories. This goes beyond just areas close to the ocean. There are areas in the United States where the chlorides in water are naturally high. If such water is utilized for mixing the stucco, over time it can lead to the degradation of the lath.

There are areas in the United States where the chloride content of the soil is high. If the stucco is carried down to ground level, soil moisture high in chlorides can leach up the wall and cause a deterioration of the lath.

Calcium chloride used to be a very common accelerant for cement mixtures. Especially in cooler times of the year, up to 2% calcium chloride would be added, so the stucco could set as fast as it would normally set during the summer, and the second coat of stucco could be added more quickly. The chloride ions would start attacking the galvanized lath. The first point of attack would be where the trowel had been dragged across the lath. This scraped some of the galvanized coating off the steel. Then if the layer of stucco tended to be thin, one could soon see the diamond shape of the lath etched in the stucco in a rusty color. The process would continue. While there are some who disagree with the concept, I believe that the lath in stucco serves the same purpose as rebar in concrete. When the lath deteriorates, and a wall is exposed to negative wind load (a partial vacuum on the downwind side of a building), the stucco can crack. If the building moves (and all buildings move), the stucco is more likely to crack than if it contained good-quality metal lath.

Protecting Metal Lath

With traditional stucco, the lath was encapsulated in the 3/8″ scratch coat. The scratch coat was then covered with a 3/8″ brown coat. The brown coat was then covered with a 1/8″ finish coat. As a result, there was at least 1/2″ of stucco cover protecting it from the environment. When one-coat stucco came onto the scene, a single 3/8″ combined scratch and brown coat was used to encapsulate the lath. This resulted in very little stucco covering the lath. The thickness of 3/8″ was selected because that was the minimum thickness necessary to fully encapsulate the expanded metal lath.

This resulted in very little coverage of stucco over the lath. While traditional stucco has always been covered with a 1/8″ finish coat or some other kind of finish coat, a finish coat was not listed in the evaluation reports that were written to allow one-coat stucco to be used. There were some applicators who felt that a finish coat was not required, and they finished their application with a whitewash of watered-down latex paint. In those areas where salt spray was common, we often saw signs of deterioration of the lath from the salt in a matter of a few weeks. For those who have forgotten basic chemistry, sea salt or table salt is sodium chloride. There are chloride ions and sodium ions. Since sodium chloride is very soluble, the chloride ions are very readily available. When we saw deterioration of the lath from salt spray, we asked the lath manufacturers how much stucco was required to protect the lath. We never got an answer. Recently this question came up in *ASTM*, and several people came up with answers; but on questioning them, I could not find a technical basis for their answers. Recently an addendum was added to *ASTM C 926* to notify design professionals that in corrosive areas the lath needs to be protected.

But it is not simply salt and other soluble chlorides that are causing the problem. There are chloride polymers. In the 1960s, a mortar and stucco additive called Sarabond was commonly utilized. It was a Saran latex; and while it had wonderful properties, with time it caused the deterioration of metal imbedded in the mortar or stucco. That included, but was not limited to, stucco lath. There are a number of people in the *ASTM* cement community who were young professionals when the Sarabond problems started surfacing. They want no similar occurrence on their watch.

Attaching the Lath

Lath needs to be attached to the frame of the building, not to sheathing. The most common standard for stud construction is to apply one fastener every 7″ on each stud. Since most construction in the United States utilizes studs spaced at 16″-on-centers, that is 112 in^2/fastener or 1.29 fasteners/ft^2. An interesting factoid is that the pullout strength of a one-inch long construction staple in construction grade lumber is about 65 pounds. At 1.29 fasteners/ft^2, we're talking about approximately 83 pounds resistance/ft^2 to pull out the minimum number of staples required to hold the lath in place. When we approach hurricane-type winds, a negative wind load on a wall can exceed 65 lb/ft^2. How many staples have to end up in the sheathing rather than in the framing for a wall to be compromised when the negative wind load is pulling on the stucco façade? I am not a registered engineer; therefore I cannot furnish you with an engineering answer to this question. What I do know is that there needs to be a fastener every 7″ on each stud and the only thing the fasteners do that end up in the sheathing is to allow water to enter where it should not be entering.

Some people go wild with a staple gun. They use the staple gun to smooth out every little dimple, including the self-furring dimples, in the lath, rather than stretching the lath out and applying a few appropriately-placed staples.

At a demo in College Station, Texas, some years ago with some building inspectors present, we had five 2'×4' panels that we had prefabricated for demo purposes using self-furred lath. One of the local contractors was going to be demonstrating applying stucco. Before he started, he got out his staple gun and added staples. In one panel I counted over 60 staples in 1 ft^2. When I stopped him and talked about the problems with extra staples, he stated he had been applying stucco for 20-some years, and this was the way it was done. One of the building inspectors said that the nailing pattern was just the minimum and that he encouraged the addition of extra staples. When I pointed out that the lath was not furred from the WRB, another inspector said, "It meets code. The lath is self-furred."

Editor's Note: The code speaks of nailing pattern but allows staples to be used as well.

ASTM has standards concerning how far out the lath must stand to meet the self-furred criteria. If it is stapled to the studs, and not in the field area, it usually hangs on the wall in an appropriate furred manner. As soon as the lather starts adding extra staples, the self-furring is defeated.

Many code inspectors accept the self-furred lath as meeting the furring requirement. The good ones mandate that the lath be on the wall in a furred manner.

Separate and apart from flattening the lath, extra nails and staples can also lead to water penetration problems. Each penetration through the WRB increases the possibility of water passing through the WRB. Two staples/ft^2 probably will not allow a significant amount of water through the staple holes, but how much water will 60 staples/ft^2 allow to pass through the WRB?

Many kinds of sheathing are used. When the sheathing is gypsum board, or EPS board, nails and staples have nothing to grab hold of. Many times they can be pulled out by hand. Inserting nails or staples in the sheathing serves only to degrade the WRB. If we go back to the 60 staples in 1 ft^2, and do a little calculating, we can surprise ourselves. If the average leg of a staple damages an area one-eighth-of-an-inch by 1/4", then the 120 legs of those 60 staples would have damaged nearly 2 in^2 of that 1 ft^2 of WRB. Do you think carving out a 2 in^2 hole in the WRB would be acceptable?

Remember, if the lath is not furred-out, it is at the back of the stucco panel and cannot provide as much protection against cracking. If extra staples are used, there is a possibility that water will pass through the WRB.

Chapter 11
Lath Accessories

Lath is just relatively-flat pieces of material. If we did not have accessories, stucco would not look as sharp as it does. Casing bead and control joints serve as screed points so panels of stucco are flat. Corner bead or corner aid ensures that corners are square, and not rounded. Without the various accessories, a stucco job would probably look fairly pathetic. In this chapter we will go over the different accessories and explain their uses.

Casing Bead

Casing bead is sometimes called J channel. If you look at it in cross-section, it looks like a lowercase J. It is the basic accessory to end a stucco run.

Traditionally, the casing bead was added to the wall prior to lath being placed on the wall. This changed with the advent of paperback lath. If the casing bead was installed along with control joints and weep screeds, and then the lath was cut and installed, there would be no continuous WRB. So some plasterers started installing the lath first, and then nailing the casing bead and other

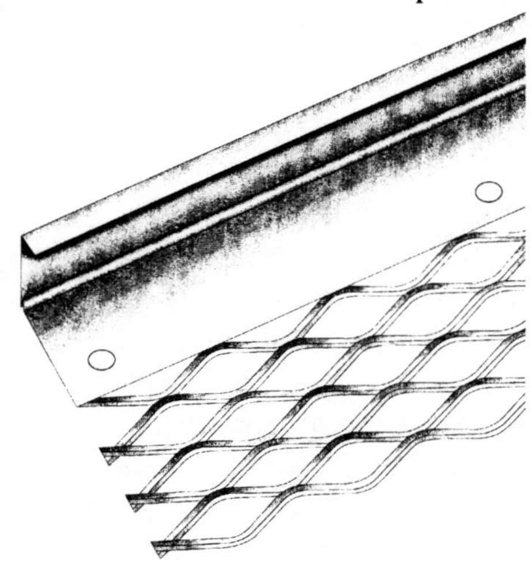

Figure 7. Casing bead

accessories on top. There is now an entire generation of lathers who have never seen the casing bead installed first. They can argue with conviction that installing the lath first is the way it's always been done. With the way the lath is lapped and the paper backing is lapped, it is very easy, utilizing this modern shortcut, to end up with a scalloped edge of the stucco.

Casing bead is used at the top of the stucco run against the sides of windows and doors, and at the bottom of windows. Casing bead comes in a number of different widths. The most common widths are 3/8″ and 1/2″ for one-coat stucco and 3/4″ for three-coat stucco. In cases where EPS is placed behind the stucco and the lath, larger casing bead may be used that encapsulates both the EPS thickness and the stucco. Traditionally, casing bead was the first accessory attached to a wall. It was placed far enough away from window frames and door frames so that backer rod could be installed between the frame and the casing bead.

Weep Screed

Weep screed is a fairly new accessory. Up until 20 years ago, the bottom of stucco panels normally ended with casing bead, but to enhance drainage of the wall, weep screed started being required. The simplest weep screed is a casing bead with holes along the bottom edge, plus flashing installed behind the WRB and bend-

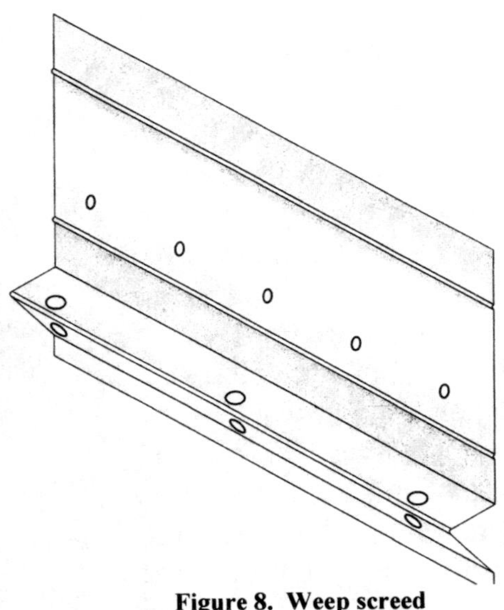

Figure 8. Weep screed

ing out so that any water discharge from behind this stucco will not run down the wall. There are other forms of weep screed that have the flashing built in and that do not end with the typical J of the casing bead, but with a sloped surface so that water that might be behind the stucco can easily exit.

A problem with installing weep screed is that seldom is a foundation perfectly square. The framing on top of the foundation, therefore, does not line up with the foundation. Sometimes due to sloppy workmanship, spaces are left so that various critters can enter the wall behind the weep screed. Many contractors, to prevent the entry of critters into the wall, caulk the entire area. This prevents water from exiting, and completely negates the weep screed. If there is a 4-inch gap between the foundation and the weep screed, and I have seen a number of 4-inch gaps over the years, who is at fault? Is it the lather, the framer, or the person who poured the foundation? Even though the lather normally gets the blame, much of the blame should probably go to the contractor who did not run a string-line along the forms before the concrete foundation was poured and the builder who was more concerned with speed than quality.

Control Joints

Control joints and expansion joints are regularly confused. Expansion joints are used where there is an expansion joint through the building. Control joints are used where there is thermal or other

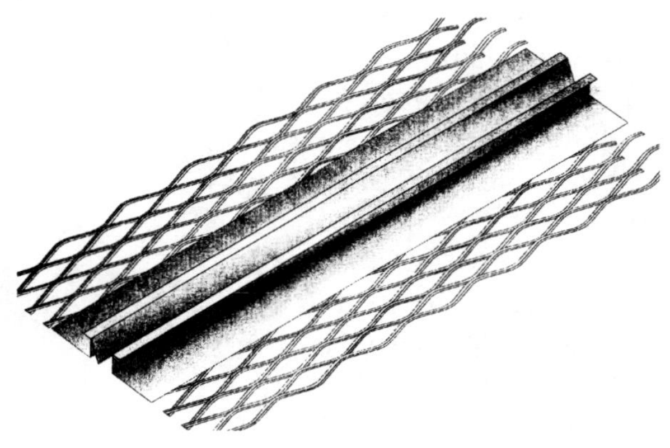

Figure 9. Control joint formed with two casing beads back to back

movement of the façade.

Expansion joints are most commonly used when different sections of a building are expected to react to movement in a different manner, and especially where clay masonry is used to construct a building. Fired clay products are smallest when they come out of the kiln and increase in size after that. The expansion joint provides room for that expansion. These joints allow three-dimensional movement. Cement products, including stucco are at their largest size when poured or applied. After that, the cement products shrink. If accommodations are not made for the shrinkage, cracks will occur. Most larger buildings have control joints and/or expansion joints if you don't believe me, look behind the downspouts that drain water off the roof, or look behind the cemetery trees. Cemetery trees are tall, slender arborvitaes that do a wonderful job camouflaging joints. Control joints usually allow movement in only two directions. If the joint goes through a wall, it is an expansion joint. If it is only on the façade, it is a control joint.

It wasn't too many years ago that we did not have the nice, preformed control joints. What a plasterer did was to cut a groove with his trowel in the scratch coat. This would provide a weak spot in the wall where the stucco could crack. Then the brown coat was installed over the scratch coat. The crack in the scratch coat would crack the brown coat.

The most common control joint looks like the letter M, with two horizontal flanges leading from it. These flanges are expanded metal. As a result, it's easy to tie them to the lath.

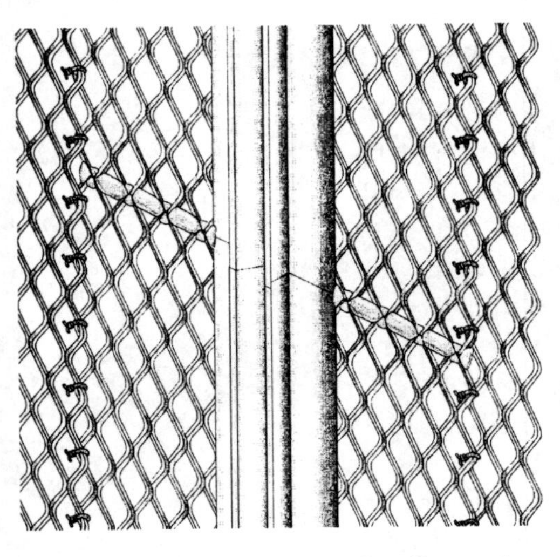

Figure 10. Use a bead of caulk at each

From time to time, some lathers form control joints and expansion joints from two casing beads that are positioned back to back, which is a perfectly good solution.

There are horizontal control joints and vertical control joints. In general, the horizontal control joints go at every floor plate. Vertical control joints go down from the corners of windows to the next horizontal control joint, and go up from the corners of windows and doors to the next higher horizontal control joint. Other vertical control joints are added so that the panel size complies with the requirements of *ASTM C 1063*.

Water has a tendency to run down vertical control joints; therefore, vertical control joints should not be interrupted. As a result, vertical control joints should be installed before the horizontal control joints. In installing vertical control joints, you want as few joints as possible. If you must join two control joints, use caulk behind them. While the illustration shows the critical point for the caulk, I have never seen one done that neatly or with that small amount of caulk.

Care should be taken when a horizontal control joint intersects a vertical control joint. The flange on the end of the horizontal control joint should be trimmed back so that the screed point on both control joints is the same height. The standing portion of the horizontal control joint should be kept 3/8″ to 1/4″ away from the standing portion of the vertical control joint. If it is not, the horizontal control joint may interfere with the action of the vertical control joint, and there may be some spalling at the interface.

Control joints function by expanding or contracting with movements of the stucco. If they are nailed solidly to the framing of the building, they cannot expand or contract. They become screed points. If the control joints are nailed solidly in place, often cracks will occur in the stucco close to the control joints. Control joints normally come with plastic tape over the groove in the top of the letter "M." The tape is there to keep stucco out of the groove. If you get stucco in the groove, clean it out immediately. The easiest way to do this is to ensure that the tape remains in place. If that doesn't happen, use the edge of the trowel to clear out the stucco that's in the groove, and then add a rag to the blade of the trowel, and wipe the groove down.

When the wall is finished and you have someone on site to seal the windows, tell him that he should under no circumstances add caulk or sealant to the grooves in the control joints.

Expansion Joints

Control joints provide for two-dimensional movement of the façade of the building. Expansion joints provide for three-dimensional movement of the building. Wherever there is an expansion joint in the building, an expansion joint should be added to the stucco. If the area is lathed over and ignored, a very nasty-looking crack will develop. The simplest expansion joint to construct is to end each stucco panel before you get to the expansion joint. End each panel with casing bead. Leave at least half an inch of space between the casing beads, and insert a backer rod into the space. Then the gap should be sealed with the most long-lasting and elastic sealant you can find.

If you decide to install a control joint over the expansion joint in the building, it might work; but I would be willing to bet that a nasty-looking crack will develop. How do you explain to the building owner, and his attorney, that they have to accept a nasty-looking crack, so that you can save $200 on that million-dollar building? It might be a little bit of an exaggeration, but you get the point.

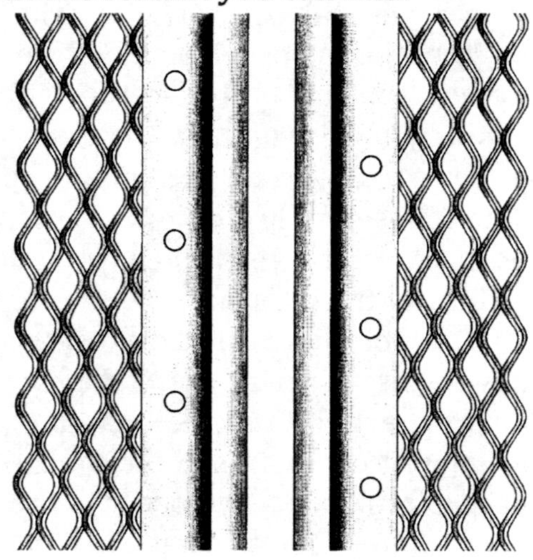

Figure 11. An expansion joint formed with casing bead

What is really interesting is when adhered concrete masonry veneer is applied across an expansion joint. There is no way that the customer will be pleased a year after the installation is complete. But it is done from time to time because the applicators and the design professionals either do not know the problems that can develop or they are too busy to handle the details.

Most expansion joints are camouflaged with downspouts, vegetation, window mullions, or quoins at corners.

Corner Aids

Not all stucco façades are in the same plane. There are outside corners. There are inside corners. There are recessed window and door frames. Each of these has a potential of cracking where the plane of the stucco changes.

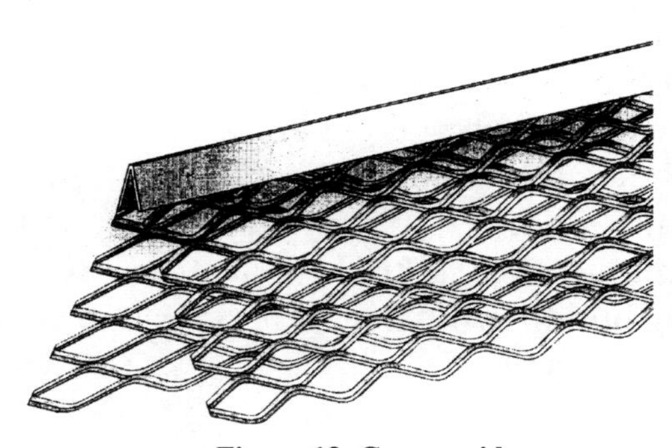

Figure 12. Corner aid

There are ways to prevent this cracking from occurring. The first method, and this should be used in every installation, is to not stop a piece of lath at a corner. The lath should extend at least 16″ beyond the corner. This will allow the lath to be nailed to a stud, if studs are being used. The next piece of lath should start at the stud on the other side of the corner. This provides two pieces of lath around the corner and it substantially reduces cracking at the corner. This technique should be used for both inside corners and outside corners.

For outside corners, the aesthetics of the job will be much higher if you have a screed point that is lined up with the plane of each of the panels forming the corner. The lath accessory for doing this is called a corner aid, or a corner bead. These devices do not come in standard thicknesses like casing bead, control joints, and weep screed. The lather bends the corner aid and staples it into position. Staples are driven into the framing lumber at the corner and are usually stapled every 7″. While I have seen lathers do this rapidly, I always need somebody else to help me hold it, and it takes forever to get it positioned correctly. As my friend Ricci says, "The hands have to learn, as well as the brain." Apparently, my hands have not learned.

For inside corners, it is appropriate to pre-bend the lath so it fits tightly into the corner. After it's been nailed in place, some lathers place a 2 × 4 in the corner, and tap it with a hammer, to ensure that the corner is uniform and sharp-looking from top to bottom.

On more than one occasion while examining a building with sharp-looking inside corners, I have noticed cracks. Inside the building there are signs of water. In taking the system apart, we find that paperback lath was used. In order to produce a sharp-looking corner, that paperback lath was cut at the corner. That means there is no WRB at the corner. There is no continuous lath at the corner. With building movement, a crack must develop. With no continuous WRB, there is no way that water is going to stay out of the building. Most jurisdictions require a lath inspection by code officials. If they weren't specifically looking for this problem, there's no one else around the job site who would notice it.

As an added precaution, to protect both inside and outside corners, I like to see an extra layer of WRB at the corners.

Chapter 12
Principles of Formulation

At this point I am going to disappoint at least one person who's reading this book. While I spent a portion of my career developing formulae for stuccos that were manufactured and placed into bags, this chapter, and this book, are not going to go into developing that kind of formulae. This chapter and the next several chapters are going to go into mixing stucco on the job site from various components. If you were expecting the secrets of developing bagged stucco that the man with the trowel in his hand loves, I would strongly recommend that you read the entire series of ***The Stucco Book***, get your PhD in cement chemistry, go out and apprentice yourself to a plasterer for 4 years, and then start playing in the lab.

We could start off this discussion by jumping in and stating that a person needs to put 2/3 to 3/4 of the mixing water in the mixer and then add half of the sand. Such a discussion would not contribute to the reader understanding the interrelationships among the components that are mixed on the job site.

Abram's Law

Most people who deal with concrete understand, or at least are familiar with, the concept of water/cement ratio. The more water added to concrete, the weaker the concrete will be. Abram's Law simply places that concept into a mathematical formula. When looking at the same components, the same curing conditions, and the same testing procedures, the strength of concrete is inversely related to the water/cement ratio that is used. After the water needs of the components are satisfied the lower the water/cement ratio, the higher the compressive strength.

Many sources do not mention that Abram's Law is a simplification of Feret's Law, and most of the sources do not give the mathematical formula for Abram's Law. In keeping with this long-standing tradition, I will not, either; but if you want the formula, you can develop it from the formula for Feret's Law which is listed in the next section.

Feret's Law

Feret's Law states that there is a relationship between the absolute volume of cement, water, and air that can be used to predict compressive strength. When Feret developed the law in the late 19th century, he was considering compaction, or lack thereof, as the source of air, rather than entrained or entrapped air. We have found that Feret's Law is applicable when entrained air or entrapped air is considered.

Some years ago I referred to this law in a draft peer-reviewed paper concerning mortars and stuccos. The editor refused to allow its use; because there were not any peer-reviewed papers that had stated that this law was applicable to mortars and stuccos. In my own unpublished testing, I have found a good correlation.

Feret's Law can be written as

$$S = K [c/(c + w + a)]^2$$

where

S	Compressive Strength,
K	Constant derived for the components used,
c	The absolute volume of cement,
w	The absolute volume of water,
a	The absolute volume of air.

As you can see, the absolute volume of air affects the compressive strength of a sample in the same manner as the absolute volume of water.

There are many admixtures that work as water-reducing agents, but do so by entraining and/or entrapping air. Probably the most common one is liquid detergent. You get by with less water in the mix, you pat yourself on the back for reducing your water/cement ratio, and you have reduced the compressive strength that you thought you were increasing. This holds true whether you're making concrete to pour a foundation, or you're making stucco to apply to a wall. Just because somebody told you to add something to your mixer, don't do it, unless you know what the result is going to be.

Since I brought up liquid detergent, I might as well mention that a few years ago in an *ASTM* committee meeting concerning masonry cement, I made the statement that there were many people adding Ivory Liquid as they job-site mixed their masonry cement. One of my friends, a cement chemist with a major cement company, stood up and stated that such an atrocity should never be allowed to happen. He went on to say that all knowledgeable people know that they should be using Dawn Detergent. He did it all with a completely-straight face. I stood up and stated that I was from San Antonio, and that many masons in San Antonio did not realize that what they were doing was wrong, and that Ivory Liquid was not as good as Dawn Detergent. As solemn as most of us remained during this discussion, there were some newcomers in the room who didn't realize we were all laughing inside.

Nordmeyer's Law

After working with water and air in mortars and stuccos, and adding other fine components, I developed Nordmeyer's Law. This is the first time it is being published, so this is an historic occasion. In simplest terms, Nordmeyer's Law states that dust, any nonreactive component finer than 200-mesh, affects compressive strength in the same way as the absolute volume of water and the absolute volume of air.

Nordmeyer's Law can be written as

$S = K [c/(c + w + a + d)]^2$

where

S	Compressive strength,
K	Constant derived for the components used,
c	The absolute volume of cement,
w	The absolute volume of water,
a	The absolute volume of air,
d	The absolute volume of dust.

How does this affect mixing stucco on the job site? I run across people who use unwashed sand that contains fines that are finer than 200-mesh. I run across people who add fly ash on the job site, and they add more fly ash when there are calcium ions to react with it. This excess fly ash is just so much dust. There are a lot of people who add pigment to their stucco. Most pigments for light colors are added at about 2% of the concentration of the cement, but for darker colors, oftentimes as much as 8% of the concentration of the cement is used. If the sand carries 5% fines, there is 5% extra fly ash, there is 8% pigment, and a squirt of the Ivory Liquid adds an extra 15% air content; we have a very creamy mix that cures as if we had added 33% extra water to the mix.

Chapter 13
Cements For Stucco

Now we get to the first of the components that go into job-site mixed stucco. Cement is cement, but there are several kinds of cement, and most of those kinds of cement are further subdivided into Types.

Portland cement includes Types I, II, III, IV, and V. Blended hydraulic cement includes Portland blast-furnace slag cement and Portland-pozzolan cement, to name two. You probably will not find all of the possible kinds and types of cement in any one market, but this section will serve as a guide to selecting appropriate cement to use for producing stucco. The first division will be those cements that are mixed with hydrated lime to form stucco. The second division will be those cements that do not need to be mixed with hydrated lime to form stucco. They may already contain hydrated lime, or more likely, they contain a plasticizing agent that gives the material body.

Cements Requiring a Plasticizer

Portland Cement—ASTM C 150

This is the workhorse for structural purposes.

Type I—for general use

Type II—moderate sulfate resistant and moderated heat of hydration

Type III—high early strength

There are air-entrained versions of the above types.

Most of the Portland cement produced in the United States meets both the Type I standard and the Type II standard and is designated Type I/II. Type IV and V Portland cement exists, but is not commonly specified for use in stucco.

Blended Hydraulic Cement—ASTM C 595

These cements are considered greener than the Portland cements, since they often contain fly ash or ground granulated blast-furnace slag, both of which are recycled products. They work differently than Portland cements, but produce equal results when the short learning curve for their use is mastered.

Type IP—for general use when a pozzolan is desired

Type IS—for general use when a slag is desired

There are air-entrained versions of the above types.

ASTM C 595 has recently been updated, but *ASTM C 926* has not yet been updated to show the new type designations for blended hydraulic cement.

Hydraulic Cement—ASTM C 1157

These cements meet many of the same physical requirements as Portland cement, but do not have the chemical composition requirements that Portland cement has. While they can probably safely be used for stucco, they are not currently listed as acceptable in *ASTM C 926*. They are included in *ASTM C 270* so they can be used for mortar.

Type GU—hydraulic cement for general use. Type I Portland cement will fall into this classification.

Cements That Don't Need a Plasticizer

Masonry Cement—ASTM C 91

Masonry cements are designed for laying brick and block, but are commonly used for stuccoing as well. While many masonry cements exhibit good bond strength, the standard does not have any bond strength requirements. Bond strength is an important factor for many, but not all, applications of stucco. These include, but are not limited to:

When stucco is direct applied to concrete,

When stucco is direct applied to masonry,

When a brown coat is applied to the scratch coat,

When adhered concrete masonry veneer is attached to a scratch coat.

The following types of masonry cement are produced and used for stucco:

Type N—Usually when Type N masonry cement is used, an extra shovel of Portland cement is added to increase the strength.

Type S—This is the most common masonry cement used for stucco.

Type M—This is a higher-strength product and as a result is often harsher and harder to work with than the Type S masonry cement.

Editor's Note: While ASTM C 926 allows the use of Type N masonry cement with a 28-day laboratory compressive strength of 900 psi. ASTM C 1329, Plastic (Stucco) Cement does not have a Type N. The lowest 28-day laboratory compressive strength for this product is 2,100 psi.

Now you must endure an explanation concerning the origin of N, S, and M. When *ASTM C 91* was being developed, the types were originally identified as Types 1, 2, 3, 4, and 5. There was a discussion concerning the interpretation of those types. There were those who thought some users would think that Type 1 masonry cement was better than Type 2. As a result, the letters A, B, C, D, & E were proposed. Again, some thought that some users would think that Type A was better than Type B. Bryant Mather suggested that since we were dealing with the works of masons, we should use the word MASONWORKS. This fell on deaf ears until he added that if we take every other letter M, S, N, O, and K, we can name the types and no one can conclude that one type is better than another type based on the name. After some discussion and someone, who will remain anonymous, suggesting that forever after people would wonder about what the letters meant, the group voted to use those designations. At the time *ASTM C 91* was adopted, Types O and K were very common. Since then, there has been a move away from them.

Mortar Cement—ASTM C 1329

Mortar cements are very similar to masonry cements, but they have a bond strength requirement. Mortar cements were developed to address a prohibition on masonry cement in structural masonry in zones of high seismic risk. Since mortar cements came on the market, the bond strength of many masonry cements has improved. Mortar cements also meet the masonry cement requirements.

Type N
Type S
Type M

Plastic (Stucco) Cement—ASTM C 1328

Plastic cement is a term used on the West Coast. It is a product produced at a cement plant and is similar to masonry cement but is made specifically for use in stuccoing.

Stucco cement is a term used on the East Coast and is the same product as the plastic cement. When the standard was originally drafted, it got hung up in committee because East Coast and West Coast producers could not agree on a name. The discussion was so boring that I had to excuse myself to go outside and watch the grass grow for some excitement. A compromise was finally developed which resulted in the name "Plastic (Stucco) Cement." In the central portion of the US, the term which I coined many years ago, Scratch and Brown cement, is commonly used for the same product. This cement is available in two types.

Type S
Type M

One-Coat Stuccos

One-coat stuccos are not covered by an *ASTM* standard, but rather are covered by evaluation reports. Some years ago there were nine manufacturers of one-coat stuccos who were members of *ASTM Committee C11*. I was the eighth one to draft an *ASTM* standard for one-coat stuccos. It survived in the task group about as long as the previous seven drafts. Each manufacturer had opinions as to the parameters that were important, and there was not much agreement among the members of the task group. The original one-coat stuccos contained acrylic polymers and, as a result, were expensive to produce. Other manufacturers found ways to produce a product that did not contain the acrylic polymers. Those who used acrylic polymers wanted to mandate that all one-coat stuccos contain a minimum amount of acrylic polymer. Those who had been able to formulate away from the acrylic polymers objected. Those were interesting times. We butted heads and got nowhere.

Chapter 14
Sand

Everybody knows that sand is sand. It's an inert material that is used to extend the amount of stucco or mortar that a bag of cement can produce. At least there are people who believe that. But not all sand is created equal. Some sand is finer than other sand. Some sands have a round particle shape, and other sands have an angular particle shape. Some sands are porous, and other sands are not porous. Some sands are lightweight, and some sands are heavyweight. Among the lightweight sands are pumice, perlite, and lightweight aggregate. Among the heavyweight sands are basalt sands and barite sands. We will discuss each of these variations.

If you call a sand processor and ask for a truckload of sand, first thing he's going to ask is what type of sand you want. He may have several types available, such as sandblasting sand in several different size categories, casting sand, casting, concrete sand, masonry sand, stucco sand, and blow sand.

If you go to a building CODE book, or an *ASTM* book of standards, and you look up sand in the index so that you can figure out what kind of sand you need, you may find that the index does not list any sand. What all the rest of us call sand, they call fine aggregate.

Gradation

Each of these sands will have a different gradation. Let's start with blow sand; this is very fine sand that is often used between paving units. The concept is that if the tiny gaps between the units are filled, the units will stay in place better. If you have sand particles that are too large, they won't make it down into those tiny cracks. Most of the large particles of blow sand are less than 1/32″ in diameter. Many of the fine particles are smaller than 200-mesh (Particle sizes are often determined by passing the material through standard sieves. A 200-mesh sieve has 200 wires per inch in each direction. The space between the wires is 75 microns—0.003 inches. We can eliminate blow sand from sands for making stucco, because with those fine particles, additional water is required to make a workable mix and the fine particles do not contribute to strength development of the cement paste. This results in lower strengths.

The finer a substance, the larger the surface area. Consider a cube that is 1″ on edge. The surface area is 6in^2. If you chopped it up into 1/2″ cubes, each cube would have a surface area of 1 1/2 in^2, but you would have 8 cubes. This would give you a total surface area of 12 in^2. It works with spheres also, but the math is a little more complex. If you halve the diameters of the particles in a given weight of a substance, you double the surface area and double the area that needs to be wetted. That is one of the reasons that fine sands hold more moisture and require more moisture to become wetted.

Many people who carve stucco like to use 60-mesh (250 micron or about 0.01″) sand in their carving mix. 60-mesh sand, without a bunch of fines mixed in, does not require as much water as blow sand, and it has no large sand particles which will create a rough surface when carving. Since most of the sand particles are about the same size, extra cement is required to fill the voids between the sand particles.

Grades of sandblasting sand contain a narrow size range. If other things were equal, one could mix several grades of sandblasting sand and come up with an appropriate gradation for stucco sand. Stucco made with an excellent blend of manufactured sandblasting sand would be harsh when troweled. That having been said, since sandblasting sand is dry, many manufacturers who make sanded stuccos, and do not own their own sand-processing plants, will use a mixture of several grades of sandblasting sand.

Concrete sand is, of course, designed to work with concrete. The gradation needs to be such that strength can be built, and it needs to be able to be finished. As a result, it can have up to 10% passing through the 100-mesh sieve for workability, and up to 5% can be retained on the 4-mesh sieve. By having a range of particle sizes, greater strength can be built.

Masonry sand is designed for making mortar for laying brick, block, stone, and other masonry units. Since the joints are often 3/8″thick, maximum size has to be limited. As a result, with masonry sand, all of it needs to pass through a 4-mesh screen. To increase the workability with the trowel and to allow the surface of the mortar to be tooled with only one stroke of a mortar tool, up to 5% of the sand can be fine enough to pass through the 200-mesh sieve.

Stucco sand, as the name implies, is designed for stucco. Since finishing stucco normally does not utilize as many trowel strokes to bring fines to the surface as finishing concrete does, but it requires more tooling than mortar joints, the stucco sand normally carries more fines than concrete sand, but not as much as masonry sand. Since stucco is often applied in 3/8″ layers, all of the stucco sand needs to pass through a 4-mesh sieve.

Probably the biggest difference in the standards for the three sands is that concrete sand and masonry sand are measured based on the amount of material retained on each of the test sieves, and stucco sand is measured based on the amount of material passing through each of the test sieves. Some sands will meet all three gradations. Each standard also includes a caveat that states if the sand does not meet the gradation, but there is a history of its being used successfully, it can continue to be used. Of course there are also a few weasel words in the standards.

Over the years I have seen numerous people pick up a handful of sand and say this is stucco sand, or this is masonry sand. They must have a better eye for the sand than I do, because the only way that I can tell whether sand meets the gradation is to run the tests. I can also mix up a batch of stucco and make judgments based on workability, but picking up a handful of sand and making a pronouncement about quality of the sand is wishful thinking.

Sand needs to be graded so that the smaller particles can fit between the larger particles and leave as little space for cement particles as possible. If we had stucco sand where a majority of the particles were between 4- and 8-mesh, and there were some particles that were finer than 100-mesh, in order to make a workable mix it would be necessary to add a great deal more cement to fill up that space between the particles. *Section 6.1.1* of *ASTM* C 897 states that not more than 50% shall be retained between two consecutive sieves, and not more than 25% between the 50-mesh and the 100-mesh sieves. These additional requirements ensure that the stucco mix is workable without adding extra cement.

Particle Shape

Manufactured sand is sand that is produced by crushing larger aggregate. Depending on the type of crusher used, and the type of stone used, the particle shape can range from slivers to cubes. Natural sand normally has round edges. In some cases the particles may be nearly spherical; in other cases the edges of the particles have just been softened. One of the most common sources of natural sand is from riverbeds or from former riverbeds. Another source of natural sand is ancient beaches. Water washing the sand tends to wear the sharp edges off the sand particles. Normally with a particular type of rock, agate for example, the more rounded the particles, the less bond strength there will be between the particles. When you move to a different type of rock, chert for example, the absorption of the rock will be different, therefore the bonding to the rock will be different for the same shape particle.

Density or Unit Weight

There is confusion between the terms density, unit weight, bulk density, and specific gravity. We will use the system I learned in the early 1950s. What I call density; younger people may call unit weight or bulk density. What I call specific gravity, some may call density.

Sand density is caused by the specific gravity of the sand, by the gradation, by the voids within the sand particles, and by whether those voids can absorb water. Sand density is commonly measured as pounds/ ft^3. A confusing issue is that sand is not measured in its dry and compacted state like most other substances, but it is measured in a loose, damp state.

Specific gravity is the weight of a material in relationship to water. To confuse the matter, there is actual specific gravity and apparent specific gravity. A ball might be made out of polyethylene plastic and would weigh 1.05 times as much as an equal amount of water so the specific gravity would be 1.05. If the ball were hollow, it might weigh 10% of the amount of an equal volume of water, so the apparent specific gravity would be 0.10. We do not have hollow sand particles that cause sand to have that much difference in specific gravity and apparent specific gravity, but we do have pumice sand and perlite. Pumice is a lava-type rock with air bubbles in it. Pumice may be light enough to float. To get smaller particles of pumice, it must be milled, and many of those bubbles are broken. As a result, the apparent specific gravity of the pumice increases as the particle size decreases. Perlite is a volcanic glass that can be expanded when it is heated. It pops, kind of like popcorn. Many of the particles are spheres which do not allow water to enter. When you need another gradation of perlite, the material is screened or air-separated. It is not milled. The density of sand is decreased by materials in the mix that contain voids, like the perlite, pumice, and lightweight aggregate.

Barite has a specific gravity of 4.5, so it is four-and-a-half times heavier than water. Sand made with barite, or basalt, another heavy material, is heavier than normal sands. Products made with barite sands are often used as radiation shields. That is not something we will get into in this book.

Gradation of the sand can also increase the unit weight, bulk density, or as I call it, density. Picture a quart jar of 1/2″ marbles. You can only place a certain number of marbles in the jar. Now, pour them out and mix some 1/8″BBs. You can now put a greater weight of marbles in the jar. The gradation requirements for sand, to some extent, are an attempt to increase the density of the sand.

Absorption and Porosity

Some sand particles have voids in them, and when in contact with water, water fills the voids. Other sand particles do not have voids, so in contact with water, only the surface is wetted. As a result, it takes more water to wet some sands than others.

Loose, Damp Sand

As mentioned above, sand is measured in loose, damp condition. *ASTM* has determined that most “normal” sands that are used for stucco and for mortar weigh about 80 lb/ft^3. That is the weight of the sand, not the combined weight of the sand and the water. Since the sand carries 6% to 8% moisture, the combined weight would be about 85 to 87 pounds. The surface tension of the damp sand holds the particles apart. If the sand were allowed to dry out, the particles would fit closer together, and one could put about 100 to 105 pounds of sand in a 1 ft^3 container. If the sand were saturated, the surface tension would be lost and the particles would fit closer together. As a result, if the saturated sand that would fit into a 1 ft^3 container were dried, it would weigh about 100 to 105 lb/ft^3. When you have nothing better to do, fill a 5-gallon bucket with dry sand. Add just under a gallon of water to it and cover it. Let is sit in the sun for a couple of days and then ask a teenager who knows everything to transfer it to another 5-gallon bucket. It will not fit.

Many people wonder why sand is measured in loose, damp condition. If sand is dry, when it is wetted, it often holds onto air bubbles. If you remember Feret's Law, the more air in a mix, the weaker it is, so those air bubbles weaken the stucco. Further, as you are troweling a wall, those air bubbles may mar the surface of the stucco. While sanded stucco mixes are very convenient, they bring their own set of problems to a stucco job.

Chapter 15
Other Components

Lime

Lime does not have as many kinds and types as cement. There are a number of choices, but very few that really work well. Lime can have a number of definitions. It can refer to a calcium hydroxide product, but it can also refer to a calcium carbonate product (ground limestone). Ground limestone does not provide any value to cement stucco other than as filler; however, it has many uses in products other than stucco. Those uses include road base stabilization, soil amendments for agriculture, treatment of water, production of pulp and paper, and numerous others. Lime is one of the most common industrial chemicals in the world.

Lime starts out as limestone. It may be a high-calcium limestone, or a high-magnesium (dolomitic) limestone. Dolomitic limestone contains magnesium carbonate as well as calcium carbonate. To produce a marketable lime product, the limestone is mined, reduced in size, calcined (burned in a rotary kiln to drive off most of the carbon dioxide), then hydrated to form calcium hydroxide or a combination of calcium hydroxide and magnesium hydroxide.

Type N lime is usually hydrated in a "conventional" hydrator. Type S lime is usually hydrated in a pressure hydrator. There is a reason for this. Magnesium oxide is harder to hydrate than calcium oxide. By hydrating under pressure, more of the oxides end up being hydrated. Lime salesmen commonly refer to this as double hydration. One lime salesman many years ago responded to a series of questions from me and told me that the double-hydrated lime had two water molecules attached to it and the single-hydrated lime only had one molecule attached to it. He could produce answers, even if they were chemically impossible.

In *ASTM C 206* and *ASTM C 207*, the major difference between Type N and Type S lime is the percentage of unhydrated oxides. With Type S, the limit is 8%. With Type N, there is no limit. Since magnesium oxide is harder to hydrate than calcium oxide, if the pressure hydrator were not used, it would not be possible to meet that limit of 8%. With the pressure hydrator, all of the calcium oxide is hydrated. With the conventional hydrator, there is often 8 to 10% unhydrated calcium oxide. When the calcium oxide comes in contact with moist skin, it can hydrate and produce a chemical burn on the skin. Magnesium oxide, being harder to hydrate, does not hydrate as fast when in contact with moist skin, so it produces fewer chemical burns to the skin. Therefore, Type N hydrated limes usually burn skin more than Type S hydrated limes.

Since these products have many uses, they are produced to different standards. Some are appropriate for stuccos, and some are not. While I would prefer to tell you to use only Type S dolomitic hydrated lime when making stucco, there are people who use other limes, and I have expanded this section to include some of those other lime products.

If you are a "purist" and want to mine, calcine, and hydrate your own lime, I am willing to serve as a consultant to ensure you know the proper techniques; however, I will insist that I be paid in advance.

Quick Lime—ASTM C 5

If you want to make lime putty, but you do not want to burn your own lime, this is where you start. Lime companies have mined the lime, calcined it, and reduced it to pebble size. Quick lime is calcium oxide. When water is added to quick lime, the calcium oxide combines with water to form calcium hydroxide. In the process, heat is given off. Enough heat is given off so when quick lime was sold in paper bags, more than one warehouse burned down because the roof leaked and water got to the quick lime.

One of the labs I operated had a 5-gallon bucket of quick lime. I instructed a technician on how to safely hydrate it and warned him of the dangers of not following directions. Later he stated that I was correct. He had melted a 5-gallon plastic bucket because he chose to not follow instructions.

Many have heard of the technique of fishing with dynamite. It is illegal. When lime was regularly burned in *caleros*, people would take chunks of quick lime, place them in a quart jar along with some rocks, poke a hole in the lid and drop it in a river or stream. Fish would scatter when the jar sank to the bottom, but be moving back to their original locations within a few minutes. Water would enter the jar; the quick lime would hydrate and give off heat, which would boil the water and lead to an explosion. There were a number of things wrong with this method of fishing, including leaving broken glass in the body of water, and having most of the fish sink because their swim bladder were ruptured.

Have I convinced you yet that you do not want to hydrate quick lime? If you must, build a wooden trough with 2×6 sides. Put on safety goggles, gloves, and a Tyvek suit. Spread quick lime in the bottom of the wooden trough, and then add a little water. After the boiling and steaming stop, add a little more until you do not get any steaming when a little more water is added. This is called slaking. Traditionally, the slaked lime would be buried for 3 to 6 months to cure. If you do not want to bury the slaked lime, place it in plastic garbage cans so it can cure.

ASTM C 1489 (Standard Specification for Lime Putty for Structural Purposes) provides standards for lime putty. It specifies that the lime putty made from quick lime must be soaked for a minimum of two weeks. Talk about shortcuts degrading the industry. The longer the lime putty has been allowed to cure, the smoother it gets. My father talked about the need to cure the lime putty for at least three months, but he mentioned that his father felt that the lime putty should be cured for at least 6 months to get really smooth. In defense of *ASTM C 1489*, there are tests that are required for the lime putty that my father and my grandfather did not have.

With what you have now learned, if you think you really need lime putty, but you'd prefer to make it in a safe manner, moisten dolomitic hydrated lime and place it in a garbage can. Cover the garbage can to keep the moisture in and the air out. Leave it for two weeks to mature. Yes, there are websites that say that all you have to do is leave it for a couple hours and you have lime putty. *ASTM C 1489* lists a minimum of 20 minutes. Remember, the longer you let it cure, the smoother lime putty will be, and the longer it will function on the job site without needing to be tempered.

ASTM C 1489 states that lime putty made from Type N lime needs to be soaked for at least 16 hours. You can make lime putty from High Cal Lime or Type N Lime, but in my opinion, higher-quality lime putty is produced if Type S dolomitic hydrated lime is used.

High-Calcium Hydrated Lime

Several grades of this product are available. One grade is sold as "High Cal" Lime." It is designed for use in controlling soil pH and for other uses where particle size is not critical. If you were to use it in stucco, because it is cheaper, you would periodically have success, but occasionally you would have lime pops (conical depressions in the surface of the stucco with white chalky material in the center) show up a few months or a few years after the stucco was used. The larger particles in the "High Cal" lime are usually unhydrated chunks of calcium oxide; with time, they will hydrate. The calcium hydroxide takes up more space than the calcium oxide from which it is formed. This causes internal pressure in the stucco panel and can result in spalling on the surface.

Type N Hydrated Lime—ASTM C 206

ASTM C 206 is the Standard Specification for Finishing Hydrated Lime. ASTM C 207 is the Standard Specification for Hydrated Lime for Masonry Purposes. The two standards are often confused. ASTM C 206 provides a test to ensure that lime pops will not occur. ASTM C 207 provides for air-entrained hydrated lime and non-air-entrained hydrated lime.

Type N hydrated lime covered by another *ASTM* designation is commonly used in drinking water treatment.

Type S Hydrated Lime—ASTM C 206

Type S hydrated lime can come as dolomitic hydrated lime, or as high-calcium hydrated lime. It is possible to produce a Type S high-calcium hydrated lime with a conventional hydrator. When this is done, the lime is much closer to the unhydrated limit than it is when dolomitic hydrated lime is produced with a pressure hydrator.

Currently Type S dolomitic hydrated lime is produced in central Texas, Nevada, Utah, and the Upper Midwest. Type S high-calcium hydrated lime is produced in Alabama. With the limited production facilities, transportation has a major impact on the price.

Suppliers of ASTM C 206 Type S Lime

Lhoist

Lhoist is a European firm that purchased Chemical Lime Company several years ago. Lhoist can be contacted to determine the locations of their distributors by going to their website:
http://www.lhoist.us/frames/Frame_Plants/Frame_Plants.htm

They have dolomitic Type S hydrated lime production in central Texas, Nevada, and Utah.

Graymont Lime

Graymont Lime can be contacted to determine the locations of their distributors by going to their website: http://www.graymont.com/contact_purchase_building_lime.shtml
They have dolomitic Type S hydrated lime production in Ohio.

High-Calcium Type S Lime

Due to the high cost of bringing in Dolomitic Type S Hydrated Lime from out of state, Blue Circle Cement Company initiated production of High-Calcium Type S Hydrated Lime in Calera, Alabama, in 1998. Due to mergers, Blue Circle was required to divest itself of this facility a few years later. The facility remained in operation; however, a Google search in December, 2011, did not reveal whether the High-Calcium Type S Lime is still being produced in Calera.

Pozzolans

Why Add Pozzolans at the Job Site?

Periodically I see recommendations about adding pozzolans to stucco at the job site. These recommendations usually come from brilliant people who have never worked with pozzolans. Pozzolans work, but the amount and cost of testing needed to prevent failure does not justify the time you can save by adding them to a job-site mix.

Periodically I'm called concerning an architect who has specified that pozzolans should be added at the job site. He wants to be green and has not studied the standards and codes closely enough to understand that he is taking a risk as he tries to forge new green paths. Or maybe he does understand the risk, because another section of the specification often mandates that the quality of the stucco is the responsibility of the plasterer.

If you are complying with the building codes, and it is appropriate to comply with the codes for your area, you will not be able to add pozzolans directly to your stucco. Codes recognize the use of *ASTM C 595 Type 1P cement (Portland Pozzolan Cement)*. This product contains up to 40% pozzolan. Codes also recognize masonry cements, mortar cements, and stucco cements, which may contain pozzolans, to be used in making stucco. There are several companies that produce stochiometrically-balanced (chemically-balanced) pozzolanic cements that meet code and provide excellent workability. Two common brands that use high concentrations of Type F fly ash are Best Masonry and Magna Wall. Mineral Resources Technology produced stucco and mortar products using high concentrations of Type C fly ash. They were bought out by Cemex a few years back. A quick review of the Cemex website does not reveal that they are pushing that technology. Boral Material Technologies, which is one of the largest distributors of fly ash in the US, has been developing mortars and stuccos using high concentrations of fly ash. A sister company to Boral Material Technologies is Boral Brick, which is the largest brick producer in the US.

What is a Pozzolan?

According to *ASTM C 125 (Standard Terminology Relating to Concrete and Concrete Aggregates*), a pozzolan is "a siliceous or siliceous and aluminous material, which in itself possesses little or no cementitious value but will, in finely divided form and in the presence of moisture, chemically react with calcium hydroxide at ordinary temperatures to form compounds possessing cementitious properties."

What the definition does not say is that for the reaction to occur, the components need to be liquid, or dissolved in a liquid. That means that the silica or alumina needs to be very slightly soluble. As the soluble silica or alumina has reacted to form hydrated calcium silicates or hydrated calcium aluminates, more silica or alumina have to dissolve before the reaction can continue. That is why the pozzolanic reaction is so slow. Added to that, the cooler the ambient temperature, the less soluble the silica and alumina are.

Pozzolans are non-crystalline, so they are essentially glass.

When Portland cement molecule hydrates, a calcium ion is given off, and it reacts with a molecule of water to form calcium hydroxide. If that calcium is not tied up in an insoluble form, it will be carried to the surface by the excess mixing water. The calcium hydroxide reacts with carbon dioxide in the air and is deposited on the concrete surface as alkaline earth efflorescence. Literally, limestone is bonded to the concrete, and customers do not like that. If a pozzolan is used in an appropriate manner, it reacts with the calcium hydroxide and forms additional cement binder.

Types of Pozzolans

Class C Fly Ash
Class F Fly Ash
Metakaolin
Natural Pozzolans
Rice Hull Ash
Silica fume
Sodium bentonite
Sodium silicate
Wood ashes

Description of Each Pozzolan

Fly Ash

The most common form of pozzolan is fly ash from coal-fired power plants. Most, if not all, of such ashes contain some pozzolanic activity. Every fly ash is different. From the same source, fly ash varies from day to day and from hour to hour. Regularly I am quoted as saying that if you test a teaspoon of fly ash, it is representative of that teaspoon of fly ash. I have never made such a comment. I used the term TABLESPOON. That said, fly ash from a particular source normally maintains the same amount of variability from month to month. So people who formulate with it allow sufficient room to anticipate the swings in quality.

A few years ago, if you wanted to use fly ash, you had to be prepared to obtain it in 25-ton tanker loads. Now fly ash is bagged in 70-pound bags in some markets.

If unburned carbon is present, and there is always some unburned carbon present in fly ash, there may be a problem with obtaining an optimum level of entrained air in the stucco. Entrained air leads to workability. If extra air-entraining agent is added due to the levels of unburned carbon, and to provide sufficient air content in the stucco, a serious problem can develop if the next bag of fly ash dumped into the mixer contains less unburned carbon. Excessive entrained air leads to weak stucco.

As a byproduct from coal combustion, fly ashes are usually considered "green," but there are people who are against the burning of coal and will not give fly ash credit for being environmentally friendly.

A common way to separate Class C ash from Class F ash is whether, when mixed with water, the ash will set up like cement products do. Class C ash normally sets up, and Class F ash does not set up, or sets up very weakly. According to *ASTM C 618 (Standard Specification for Coal Fly Ash and Raw or Calcined Natural Pozzolans for Use in Concrete),* the determining criteria is the sum of the amount of silicon dioxide, aluminum oxide, and iron oxide. If the sum is greater than 70%, it is Class F. If it is between 50% and 70%, it is Class C. Usually the other main ingredient is calcium. Class C fly ashes usually contain 20% to 30% calcium, and Class F fly ashes usually contain less than 15% calcium.

Coming out of the fire ball of a coal-fired power plant are tiny drops of melted silica and alumina. In the air stream, they become spherical; and as they cool, they form microscopic spheres. The spheres that settle to the bottom of the dust collector are referred to as bottom ash. Those that float and are removed by electrostatic precipitators are referred to as fly ash.

These microscopic spheres act as very small ball bearings in any mixture made with fly ash.

Class C Fly Ash

Class C fly ash tends to set up, because the calcium is tied up with the silica and alumina in the fly ash to form low-grade cement chemical. One can describe a Class C fly ash as a blend of low-grade cement and pozzolan. Since it will set up, Class C fly ash is often used in larger quantities to replace cement than Class F fly ash. It usually sets rapidly and may contribute towards alkali efflorescence (sodium and potassium). It usually does not prevent alkaline earth efflorescence (calcium). Class C fly ashes tend to have an orange color, especially when damp.

Class F Fly Ash

The amount of Class F fly ash that can be used in a mix is the amount that will react with the calcium ions in that mix. The reactivity of the Class F fly ash can vary from day to day, and can only be determined by testing. If excess Class F fly ash is used, remember Nordmeyer's Law. The excess Class F fly ash has nothing to react with and acts as so much dust in the mix.

Class F fly ash usually is effective in controlling alkaline earth efflorescence and is not prone to flash set. Class F fly ashes tend to have a beige color.

Metakaolin

Metakaolins are produced when some clays are heated to drive off some, but not all, of the chemically-combined water. Usually they are very consistent when they are coming from the same source and tend to be light in color. The Romans produced metakaolins when they under-burned brick and then ground them up.

Natural Pozzolans

Natural pozzolans include volcanic ash, volcanic tuffs, and other volcanic materials. Much of the soil in some parts of

Mexico is pozzolanic. All the plasterer has to do is mix some lime putty with the soil and he has an excellent plaster.

A band of volcanic ash is found in Texas from Starr County to Campbellton, on towards the Louisiana line. The closer to Mexico it is, the greater the depth. This material makes an excellent pozzolan. It was used in Falcon Dam, the Galveston Sea Wall extension, Amistad Dam, and many other places. To be used, it must be mined, dried, and milled. At Falcon Dam, the specifications called for it to be calcined to remove all traces of carbon. We found the carbon source was fine mesquite tree roots that penetrated the deposit up to 60 feet. Coarser roots were removed with grizzlies (parallel bars on a steep angle used to remove oversized objects) and with screens. We also found that the fine roots did not have an impact on the reactivity of the ash, but the Bureau of Reclamation specifications included the carbon limit, so we had to comply with it.

The second big dam project we were involved in was Glen Canyon Dam. The Bureau of Reclamation identified sources of natural pozzolans, and we sampled them and started processing them in the lab. They did not mill like the volcanic ash we had used at Falcon Dam. After an initial reduction in size, we could not reduce them in size further, but the Bureau of Reclamation specifications required further size reduction. We increased our bid and lost the contract. The low bidder had not done the lab work we had done, and they were in production before they realized their mistake. The Bureau ended up pulling their performance bond. As you can see, not all natural pozzolans are alike. Shortly after that, fly ash started replacing natural pozzolans on the Bureau of Reclamation projects.

Rice Hull Ash

Some types of vegetation contain silica. When the vegetation is burned, the resulting silica may produce a good pozzolan. As a test, corn cobs have been burned in a modified coal-fired power plant. They burned well and produced a limited amount of usable Class F fly ash. Rice hull ash usually makes a good pozzolan. In its natural form the rice hulls contain opaline silica in a lace-like structure that is shaped like the rice hull. The lace-like structure provides excellent absorption. Opaline silica is a form of silica with chemically-combined water. Since it is more soluble than most forms of silica, it makes an excellent pozzolan. Two problems occur. If sufficient heat is used to burn the carbon out to comply with the *ASTM* standard, the water is driven off, and the silica is less pozzolanic. Usually companies who have rice hull ash are enamored with its value, and it becomes overpriced for use as a pozzolan. Some of this is justified, because it is an excellent absorbent. My father and I did a little work concerning burning piles of rice hulls with limestone to produce cement. We abandoned the project because we could not see a way to make a profit on it, but it appeared that in third world countries, this could be an economical way for local populations to produce cements.

Silica Fume

With all of the silicon chips that are produced, there has to be a waste material. It is silica fume. This is the condensed residue of silica vapor. It tends to be very fine, very dark, and very thirsty. Cements containing silica fume tend to be dark in color. Water-reducing agents are needed to produce a workable mix with the water/cement ratios that are used. While Class C fly ash may be used at a 40% dosage rate, silica fume is seldom used much over 5%. Because it is fine enough to be respirable (can make it all the way into the lungs when you breathe it), it is often pelletized so it is a safer product to use. In my estimation, it has no place on a stucco job site.

Sodium Bentonite

Sodium bentonite is also known as bentonite and bentonitic clay. It is made up of microscopic plates. When it gets wet, the plates separate, and the clay expands. In the process, the clay absorbs a lot of water. If care is not exercised in its use, the water/cement ratio of the mix can be overly high. Due to its expansive characteristics, it is often used to waterproof ponds.

When sodium bentonite is in the presence of calcium ions, it reacts with them and forms calcium bentonite. Calcium bentonite is more stable than sodium bentonite. At least one company produces a sodium bentonite blend and sells it as a replacement for hydrated lime. When mixed with Portland cement, sand, and water, it gives the mix enough body so it can be used as a plaster or as a mortar.

Sodium Silicate

There are a number of different sodium silicates, and the differences are related to the concentration of sodium and silica. Sodium silicate is soluble in water and probably produces more sodium ions in solution than any other chemical. Current plans call for a chapter of ***The Stucco Book—Forensics and Repair*** to be devoted to the many different uses of sodium silicate around a construction site. In the meantime, in most localities it is an acceptable practice to mix 2% to 5% sodium silicate with the mixing water for making stucco. If you add the crystals, they will not all dissolve in time, and the wall will cry (drops of water will form on the surface) about the time you have it troweled-down. By using it in the mixing water, your ultimate compressive strength will increase 10%t to 15%.

Wood Ashes

Periodically I see a suggestion to use wood ash as a pozzolan. It will work, and I have used mesquite ash successfully as a pozzolan. If a wood contains 2% silica, and 4% other ash, you have to burn a cord (4'×4'×8') (about 2 tons) to get 240 pounds of ash. Before doing this, consider how this will affect the environment. If you have wood ashes available, then consider keeping the ash dry and using it. Potassium levels in most wood ashes are high, so efflorescence will probably occur. The non-pozzolanic 4% of the ash reduces at least some of the strength gains made from the pozzolanic action.

Effect of Particle Size

In the chapter on *Sand* and the section on Gradation, the effect of particle size on the surface area was discussed. That discussion is also applicable to pozzolans. The finer a pozzolan is, the larger the surface area, and the more reactive the pozzolan is. If you halve the diameters of the particles in a given weight of pozzolan, you double the surface area and double the area where active silica can be exposed to calcium ions. This in effect doubles the pozzolanic effect.

Sources of Pozzolans

The two largest suppliers of fly ash in the United States are
Boral Materials Technologies
http://www.boralmti.com/index.html
and
Headwaters Resources
http://www.flyash.com

Both companies have a wealth of information on their websites, and both companies have fly ash in most parts of the US. Rather than show favoritism for one of the companies, I have placed their names in alphabetical order.

There are a number of metakaolin and silica fume suppliers. In recent years the industry has been consolidating, so your best way to find a local supplier is to do a Google search. Most plastering and stucco supply stores do not carry metakaolin or silica fume.

The cheapest forms of processed sodium bentonite are found in drilling mud. The cheapest forms of drilling mud do not have polymers added. Since you do not want the polymers in a stucco, check with your local drilling mud supplier and state you need the sodium bentonite drilling mud without any additives in it. Review the MSDS and the Product Data Sheet to confirm that there are no additives.

Sodium silicate is an industrial chemical. I usually purchase it from Wrico Chemical in San Antonio, since they are local for me. Check the yellow pages for Chemicals, Industrial, to find an industrial chemical supplier near you. I usually purchase sodium-meta-silicate, but any of the sodium silicates will work.

If you want to use a natural pozzolan, unless you want to mine and process the mineral, you should probably stick with metakaolin.

Mortar Fat

Mortar fats are substances that are designed to replace hydrated lime in a mortar or stucco mix. Some of them have been around for decades; others are more recent additions to the market. Some of them work very well; some of them don't. There is also a controversy as to whether or not each mortar fat is as good as the hydrated lime that it replaces. In my opinion, most of them work better if some hydrated lime is left in the mix. This section will talk around the subject, and then each plasterer or design professional can make a decision as to whether the use of mortar fat is appropriate on a particular job.

About 10 years ago one of the manufacturers of mortar fat wanted to modify *ASTM C 270 (Standard Specification for a Mortar for Unit Masonry*). He was not particularly concerned about other varieties of mortar fat, but he wanted his variety identified as hydrated lime. While this sounds like a long stretch, in his mind it was quite reasonable. As you have probably guessed, at this point you will have to endure another story.

This mortar fat has been around since the 1960s, or possibly before. It is being utilized by some manufacturers to manufacture masonry cement. It has an evaluation report and lots of documentation that state that mortar and stucco made with it are equal to mortar and stucco made with hydrated lime. The hydrated lime industry does not agree with that contention, but that's another part of the story. If mortar is made by adding the mortar fat on the job site, the mortar does not comply with *ASTM C 270.* A vast majority of mortar specified in the United States is specified as needing to meet *ASTM 270. ASTM 270* has three categories of mortar:

Cement-lime mortar,
Mortar cement mortar, and
Masonry cement mortar.

Each of those classifications of mortar can be produced in several types—O, N, S, and M.

Cement-lime mortar is made by mixing Portland or blended cement with hydrated lime, sand, and water. Masonry cement mortar is produced by mixing masonry cement meeting *ASTM C 91* with sand and water. Mortar cement mortar is produced by mixing mortar cement meeting *ASTM C 1329* with sand and water.

ASTM C 270 or *ASTM C 926* do not allow mortar fat to be added **on a job site**. Evaluation reports allow mortar fats to be used, but there is always the question whether the mortars made with them comply with *ASTM C 270* or *ASTM C 926*. Thus the request to define mortar fats as hydrated lime. *ASTM* does not modify a standard to allow one specific product to be used. Rather, ASTM standards are written for classes of products. If *ASTM* were going to act on a proposal like this, it would be necessary to write the standard so that the class or a subset of the class of products known as mortar fat could be utilized as an equivalent to the hydrated lime. The technical representative for the mortar fat company would not accept that. He wanted his mortar fat to be defined as hydrated lime. The lime industry, in protecting their turf, did not want anything except hydrated lime defined as hydrated lime. Needless to say, there was a controversy. A major cement company was prepared to launch their new mortar fat into the market, if the mortar fat company was successful. They wanted a large share of the mortar fat market.

After ten years, we decided that there was no way that a compromise between the two sides could be reached. One of the questions that had developed during those ten years, was just what parameters should a mortar meet. We knew the tests that were required for a Portland-lime mortar, but would they be appropriate for other kinds of mortar? There were lots of opinions but very few facts. Therefore, when the mortar fat task group was disbanded, a new task group was formed to develop a standard for mortars using alternate materials.

Following are a few specifics about the different mortar fats that are on the market.

Kel-Crete has been on the market for many years. While some people have maintained that this is a by-product of kelp, the name actually comes from the man who developed it, a Mr. Keller. I do not have any specifics concerning the formulation of Kel-Crete, but it appears to be a derivative of the guar bean. It only takes an ounce or two per bag of Portland cement. It comes in both liquid and powder forms. The powder is used by some manufacturers to produce masonry cements and stuccos. The liquid is used only for adding at the job site. Considering how little is used, there is always a concern about accurate measurement of the Kel-Crete at the job site. Since it serves as a plasticizer, more than one applicator has added Kel-Crete to a job site mix when he is using masonry cement or stucco cement. These products already contain a plasticizer, so adding a second plasticizer ends up producing a very weak product.

Gibco is a product that is very similar to Kel-Crete. I have talked with salespeople for both companies, and each explained to me that their company developed the formula and the other company was formed when a key employee left and started his own company. By this time, you know how much faith I place in what salespeople tell me. I find that the two products work about equally well. That means when they're misused, neither one works in an appropriate manner.

Easy Spred (that is how they spell it) comes in 7 lb bags. One bag is added to a bag of Portland cement. It contains sodium bentonite. When the sodium bentonite gets wet, it expands and gives the mortar or stucco what we call "body," like hydrated lime does. It also expands the volume of mortar, so the resulting volume is close to the amount of mortar formed when one bag of Portland cement is mixed with one bag of Type S hydrated lime. When calcium ions are given off as Portland cement hydrates, those calcium ions react with the bentonite ions to form calcium bentonite. Calcium bentonite is a more stable compound than sodium bentonite.

I've lost track of how many other mortar fats are on the market. Most of them are chemicals that need to be added to the mortar in very specific concentrations. A common dosage rate is one ounce per bag of Portland cement. How accurate do you think that dosage is going to be on a job site where the person mixing the mortar is only concerned about cashing his next paycheck? Some of them will work with Portland cement but will not work as well with blended hydraulic cement. Others seem to work equally well with Portland cement or blended hydraulic cement. If you are going to use a mortar fat, I strongly recommend that you test it with the materials you will be using on the job site before your job starts.

Chapter 16
Mixing Stucco

Mixing stucco is easy. All you have to do is rent a mortar mixer, throw some things in it, and you're done. Also, the moon is made of green cheese. In order to feed the world, all you have to do is reach up on some moonlit night, grab hold of the moon, and pull it down to earth.

Let's get serious. Mixing stucco is hard work, especially in the middle of summer. If it's not done right, the person with the trowel in her hand has to work harder. While the low person on the job is commonly assigned to the mixer, there is justification for assigning the most knowledgeable person on the job to the mixer. At least, the most knowledgeable person on the job should train the person mixing the stucco.

Probably the first thing that should be considered is ensuring that you have a mortar mixer, rather than a concrete mixer. A concrete mixer has a tub that rotates. It needs large aggregate in order to operate efficiently. It also needs a more fluid mix in order to operate efficiently. A mortar mixer has a tub that remains stationary during the mixing process, and paddles inside the tub, attached to a horizontal shaft that rotates during the mixing process. When it is time to dump a load of stucco, which is referred to as mud, the tub is rotated while the paddles are turning, and the mud flows into the wheelbarrow.

Ensure that the mixer is clean, the air filter is clean, it has fuel and oil, and it is in working order. Check to see whether the tub will rotate to dump, and will lock into place when swiveled right-side-up. If it is belt-driven, ensure that the belts tighten-up when placed into gear and loosen-up so they slip when taken out of gear. If it has a transmission, ensure that the oil level in the transmission is adequate. The paddle in the mixing tub should turn without scraping. Are all safety guards in place?

When the motor of the mortar mixer is running, never under any circumstances place your hand, a stick, a shovel, a trowel, or anything else into the mortar tub. If you need to take a sample of the stucco, either shut the machine down before taking a sample, or rotate the tub and dump the sample out. If you got a finger caught between the paddle and the tub, it would either be ground off, or smashed so it was about 1/8″ thick. Anything else stuck into the mortar mixer while it is turning will either be torn up or will stop the turning of the mixer.

Commonly, people love to fill a mortar mixer to the brim. Mortar mixers are not designed to be filled all the way. The most efficient mixing is going to occur when the shaft of the mixer is just barely covered with mud. Manufacturers understand the need to limit the amount of mud in a mixer, but they're also in the business of selling mixers, which requires marketing. Recently I was looking at a 12 ft^3 mortar mixer. It was described as a 4-bag mixer. If we look at one bag of cement containing a cubic foot, and we had 3 ft^3 of sand, the mixture normally produces about 2.7 ft^3 of mud. If we mix 4 bags of material, that gives us right at 11 ft^3 of mud. That all sounds reasonable until you get down and examine the fine print. This is the fine print that the salesman probably had never read and probably never knew existed. The manufacturer defined a bag of cement as a 50-pound bag. And the capacity of the machine was labeled as 3 1/2 to 4 bags. Four 50-pound bags of cement will produce 6 3/4 ft^3 of mud. That's just over half full for the 12 ft^3 capacity of the mixer tub. While the manufacturer called it a 4-bag mixer, while many contractors would use it as a 4-bag mixer, I contend that it is a 2-bag mixer with marketing fluff added to it.

While I'm still beating this dead horse, let me explain why this is a major irritant to me. Some years ago I had a ribbon blender that held 20 ft^3, and used it for dry-blending cement products. A ribbon blender has some of the same characteristics as a mortar mixer, but it is much more efficient. I ran an experiment to find out how long it would take to mix a batch of colored stucco. When I filled the mixer half full, in other words to the level of the horizontal shaft, mixing was complete in a minute. When I filled the blender to capacity, after thirty minutes of blending, the pigment was still not adequately blended with the other components. Considering the time it took to fill the blender, the mixing time, and the dump time involved, I ended up operating that blender with a load of 12 ft^3. That was 60% of capacity. If you fill a mortar mixer brim full, the mixing action will be compromised and you will loose workability.

The mortar mixer should be positioned close to the sand pile. Remember, the sand pile should be covered so it does not get rained on and so it does not lose moisture. That will make your job of getting a quality mix with the least amount of work much easier. You need to be able to easily shovel sand from the pile into the mixer without having to move your feet. If you are measuring the sand with 5-gallon buckets, you still need to be in that same proximity. You need a water source. Many people would use a hose with a spray nozzle on the end. This is adequate; but if you have a 55-gallon drum and a gallon container to dip water from the drum and put it into the mixer, you can more accurately measure the water you are using, and it speeds up the process. If you want to add sodium meta-silicate, you can dissolve it in your water source. If you're working on a small job, you may decide to have 5-gallon buckets filled to the appropriate level with water and use them to measure the amount of water for each batch. The dumping side of the mixer should be kept clear, so a wheelbarrow can be driven up and a batch of mud dumped in it. You may want to dig a slight depression for the wheel of the wheelbarrow so that the tub will rotate just a little further, and empty just a little more when you're dumping a batch. Finally, you need your cementitious materials and admixtures that you may be adding. These need to be convenient but out-of-the-way.

You need some tools. First, you need a wheelbarrow. Two wheelbarrows are even better. Some people think that a 2-wheeled cart works better than a wheelbarrow. Around a construction site, a 1-wheeled device is much easier to maneuver. As long as you do not fill the wheelbarrow full, it's fairly easy to maneuver. Ideally, you should have a contractor's wheelbarrow rated to hold 6 ft^3 of material. A variation of the contractor's wheelbarrow is the all-steel Mason's wheelbarrow, which many people prefer. Fill it only half full until you are comfortable using it. You need a number two square shovel. A square shovel allows you to get closer to the bottom of the sand pile without tearing up the plastic that the sand should be sitting on. A number two shovel allows a nice amount of sand to be picked up with each shovelful. A number two square shovel is pretty standard on most stucco jobs across the United States. You need 5-gallon buckets. If you have a dozen of them, you probably do not have too many. You need a one-gallon dipper to measure water. Ideally, I would like an open-top, plastic, 55-gallon drum for water. You need a hawk and a trowel. This is so you can test the mud to ensure it has the right consistency. You need gloves, a dust mask, and eye protection. Breaking bags of Portland cement and hydrated lime is dusty and the dust is dangerous if it enters your lungs. You need to protect your skin. I prefer to wear loose, but not baggy, long pants and long-sleeved shirts while working. In addition to that, you need an apron. I prefer a piece of polyethylene plastic. If mud gets on your clothing and remains damp, alkalis from the mud can work their way into your skin and cause skin burns. And last of all, you need a hat and source of drinking water, because you do not want to become dehydrated while working out in the sun.

If everything is arranged around the mixer in the order that it is to be used, there is less likelihood of forgetting to add something. Adding something in the wrong order can drastically change a batch of mud. Example: pigment is added after the sand and water are added to the mixer. The sand and water break the pigment into individual particles so it can disperse through the entire mix. If you forgot to add the pigment, and you added it after the cement had been added, you would probably have to mix a batch for thirty minutes or an hour to get the same dispersion. Mixing it that long will entrain too much air and change the characteristics of the mud. Even if you mixed it long enough so the color of the wet mud looked about right, when it dried on the wall it would be a lighter color.

Prior to this point, you should have decided on a formula that you will be using. In the previous chapters we listed the different components that are utilized to make stucco. If you don't have a reason to use a different formula, I would suggest that you start off with one bag of Portland cement, one-third of a bag of Type S dolomitic hydrated lime, and 4 ft^3 of loose, damp sand. This will give a mix that is appropriate for most scratch coat and most brown coat stucco applications.

After checking that everything is in position and is in good working order, start the mixer. Add two-thirds to three-fourths of your anticipated water needs. Additionally, you're going to be making an estimate, but after you get a little practice, you can start off with 90% of your anticipated water needs. For your first batch, it does not matter how big your mixer is, start off with one bag of Portland cement. That way you can judge how much space you have in your mixer for future batches. If you mess up by putting too much water in, you can either throw the batch away, or add additional components to salvage the batch.

Normally, stucco utilizes a mortar/cement ratio of 0.5 or 0.55. Since a bag of Portland cement weighs 94 pounds, and a third of a bag of hydrated lime weighs just under 17 pounds, we are dealing with approximately 110 pounds of cement-type material. That means we need 60 pounds of water. But there is some water in the sand. Four ft^3 of sand at 80 lb /ft^3 is 320 pounds of sand. It is going to be holding between 6% and 8% moisture. We make the assumption that it will be 8%. That way if we are wrong, we can add a little bit more water, rather than having to try to salvage a batch. Eight percent of 320 pounds is twenty-six pounds. In these calculations we are rounding for the worst circumstance. Sixty pounds minus twenty-six pounds equals thirty-four pounds. Everybody knows how much water weighs. It weighs 8.34 lb per gallon. Actually, that's the weight at about 40°F when it is at its greatest density. If we make the assumption that water weighs 8 lb per gallon, and we need thirty-two pounds of water added to the mixer, that means we need to add 4 gallons of water. Since we want to start off with 2/3 to 3/4 of the anticipated water needs, add 3 gallons of water. If your memory is like mine, write that down where you don't forget how much you added. Measure out the remaining water you anticipate using, and set it aside.

Four ft^3 of loose, damp sand are needed. Some people use 1 ft^3 containers to measure their sand. If you take 1 ft^3 of sand, the weight of the water in the sand, and the weight of the box, you have right at 100 pounds to pick up and dump into the mixer. Unless the box is designed with handles for two people to pick it up, it tends to be rather awkward and seldom used.

Some mortar mixers come with a 1 ft^3 box attached. You shovel sand into the box, screed off, and dump the box into the mixer. A more common method of measuring sand is to use 5-gallon buckets. Most 5-gallon plastic buckets will hold 5 1/2 gallons when full and level. Here is another number for you to memorize. One ft^3 contains 7.48 gallons. Just remember 7 1/2 gallons. That is adequate for what we are doing. That means 3 ft^3 of sand will fit in four 5-gallon buckets. Since we want 4 ft^3 of sand, we need five-and-a-half 5-gallon buckets of sand. Add about half of the sand; in this case, add three 5-gallon buckets of sand. Retain the other 2 1/2 buckets of sand for adding later.

Add any admixtures that you plan on using. Add any fibers you plan on using. In this case you are not adding any admixtures or fibers, so you can go to the next step. If you had added any, you would need to mix them for about one minute.

Add 1 bag of Portland cement. The easiest way to add the Portland cement is to throw the bag up on the mixer. Most mixers have a bag breaking tool on the top grate. Standing on the upwind side of the mixer, lift one side of the bag and allow the contents to pour into the mixer. After the dust clears, lift the other side of the bag and allow the contents to pour into the mixer. After the dust clears, look into the mixer to ensure that the sand, water, and Portland cement mix remains fluid. If it is not very fluid, something probably was not understood in the previous instructions.

Add 1/3 of the bag of Type S dolomitic hydrated lime. Hydrated lime comes in 50-pound bags that contain 1 1/4 ft^3. The easiest way to measure out a third of a bag of hydrated lime is to pour the contents of a bag into three 5-gallon buckets. Set 1 bucket close to the mixer to use, and set the other 2 buckets away from the mixer, so you or a helper do not inadvertently add extra lime to a batch. Do not ask me how I learned this. Add the lime to the mix; and after the dust clears, check to ensure that the mix remains fluid. It should have thickened-up at this point and be more of a paste than a fluid. Slowly add sand that was set aside for the mix. As the mix thickens-up, add some of the water that was set aside for the mix. After you have all of the sand added to the mixer, and all but about a quarter of the water added to the mixer, allow the mixer to mix for four minutes.

At this time, we will do our first test on the batch. With the mixer running, and a wheelbarrow in place, dump a small amount of mud into the wheelbarrow. Return the mixer to its upright and locked position, then pick up the mud with the trowel, and place it on a hawk. Slide the mud onto the face of the trowel and hold the trowel horizontal. Ideally, the mud should form a pile that is about 2″ high. If the mud flows off the edge of the trowel, it is too wet. If the mud shows cracking, it is too dry. If it passes the first test, then slowly tip the trowel. If the mud slips off the trowel before you get to 30° from horizontal, the mud is too wet. If it is still hanging on when you pass 45°, it is too dry.

Return the sample to the mixer; and if it was too dry, add some of the remaining quart of water, and retest. Continue doing this until the mortar passes the test. Keep track of how much water was added. Each time you add water, you should mix the mud for one additional minute. After you determine the appropriate amount of water to use, start off each batch with about 90% of that amount of water.

If the mud was too wet, you cannot use it to accurately determine the amount of water that is required for a batch. However, we can salvage the batch by adding small amounts of Portland cement or Type S hydrated lime. After salvaging the batch, you should mix a new batch, starting off with less water.

After the batch has been mixed and you're satisfied with it, place a mark on the outside of the mixer to indicate the elevation of that mix. This will help you put the right amount of sand in a mix if you end up measuring sand with a shovel. Then, with the mixer still running, look at how the mud turns over with each rotation of the paddles. Memorize what you see. That way you can look at a batch, and with a little experience, adjust to the ideal water concentration without having to pull a sample. After that, dump the contents into a wheelbarrow with the mixer running. Allow all of the mud in the mixer to discharge. Return the mortar tub to its upright and locked position.

Add water to the mixer in preparation for mixing the next batch. By this time you should determine whether you want to mix a 1-bag batch or a 2-bag batch as you continue.

As the mud is being applied to the wall, watch how it holds onto the trowel and holds onto the lath. If you think it should be just a little bit wetter, add a little water to the mud on the hawk, mix it with the trowel, and apply it. If you think it should be a little drier, just put a little Portland cement on the mud on the hawk, mix it with the trowel, and apply it.

If the person with the trowel is working on an elevated location, the mud can be shoveled up to his mortar board using a number two square shovel. If it is too high for shovel work, the mortar can be hauled up in 5-gallon buckets. If you're tying a rope to the bucket's bail, I have found that a minimum 1/2″ diameter braided polypropylene rope is more comfortable on the hands.

Many people prefer to measure sand by the shovel, and there are many people who can accurately do it. If you want to use this method, first, determine how many shovels of sand are required to fill the required number of 5-gallon buckets. Each time, try to fill the shovel to the same capacity. After determining how many shovels of sand are required, try it again and see if you come up with the same answer. When you shovel directly into the mixer, try to ensure that each shovelful is identical. Count the shovels of sand added. Also, watch that mark you placed on the outside of the mixer tub and watch the mixer shaft. Even if you were not counting the shovels, by watching the elevation of the mud in the mixer, you could add the same amount of sand each time because you are adding the same amount of cement, hydrated lime, and water to each batch.

If you are using a stucco concentrate rather than separately adding Portland cement and hydrated lime on the job site, you follow the same procedures outlined above.

If you are using a sanded mix, determining the appropriate amount of water is a little more difficult. Start off with the amount of water recommended on the bag. Add 90% of it to the mixer and retain the rest. Split a bag of pre-sanded stucco mix into at least two 5-gallon buckets. With the mixer running, add one bucket of the mix. If you are going to add pigment, it should be added at this time. As long as the mix does not get too thick, start adding the rest of the mix. If the mix thickens up, add a measured amount of water. Many bagged stucco concentrates and pre-sanded stuccos contain water-reducers and other admixtures. Some of those admixtures take a few minutes before they affect the mix. If such an admixture is in the stucco mix you are using, you may have exactly the right amount of water to give the right texture to the stucco, and after a minute or more of mixing, it may become soupy. One can work around this easily with a concentrate, but with a pre-sanded mix, you need to add a portion of the mix and let the admixture activate, and then add the remaining mix while keeping the mix on the dry side. After the remaining admixture activates, adjust the amount of water to get the right fluidity.

Now, without going back and counting, how many 5-gallon buckets were used?

Chapter 17
The Admixture Market

Historic Admixtures

Give me a little ox blood or the juice from a nopal, and I can make a good stucco mix even better, but no one follows the old ways any more. As a result, concrete admixtures are a billion-dollar-a-year industry in the US. Prior to these "essential" chemicals being developed, our predecessors used natural products. The standard formula for concrete, mortar, and stucco in many parts of the world up until the 20th century was a mixture of volcanic ash, lime putty, and aggregate. If the product was to be used as a stucco or mortar, the aggregate would be limited to sand, and the lime putty would be increased. In areas where an adequate pozzolanic source was not available, what some would consider an inferior concrete would be made with lime putty and aggregate. Here are a few of the admixtures our predecessors used.

Hair from hide-tanning operations was a common source of fiber to increase the tensile strength of the concrete. These fibers can be compared with many of the modern-day fibers in general length and diameter characteristics.

Ox blood increased the frost resistance and increased the compressive strength of concretes and plasters. Apparently it entrained a little air to increase the frost resistance, and some organic bonds developed between the fine particles to increase the compressive strength.

Lard was used to increase the water resistance of concrete and plasters. Hot water had to be used; otherwise, the lard would not disperse readily. Usually this was a summer-time operation. My father used lard when I was a kid. Lard was replaced by sodium stearate, which is soap. The sodium stearate would react with the calcium ions from the cement hydration to form calcium stearate, which is a cement binder. This would help seal the pore structure of the concrete. Traditionally soap was made by reacting lard with the leachate from wood ashes (a source of lye or sodium hydroxide). I have not researched the chemistry of adding lard to a concrete. The calcium ions probably reacted with the lard to form some form of calcium stearate. A problem with using lard was that for several years the wall would have the faint aroma of rancid lard. That is, if the aroma of rancid lard could ever be considered faint.

In Mexico, the juice of the nopal (prickly pear cactus) was added to plaster to give the mix body, so it would hang on the wall and could be used as a mortar. Often, some of the pulp would be included to serve as a fiber in the stucco.

Chopped grass or straw has long served as a fiber to increase the tensile strength of plaster and concrete and to reduce cracking during curing. Exodus Chapter 5 reports that the Israelites, while in bondage in Egypt, used straw in the production of what we would now call adobe brick. While most people have thought of the grass or straw as serving as a fiber to give the plaster, concrete, or adobe brick tensile strength, it had a more important function. Cement products and clays shrink as they dry or cure. If the center remains fully hydrated and the exterior contains less moisture, cracks will result to ease the tension. One of the main reasons for adding chopped grass or straw to a mix was to produce conduits so moisture from the center could find its way out and the material could dry evenly. While drying machine-made clay brick in a heated dryer tunnel in the 1950s, we maintained very high humidity as we raised the temperature, so the outside of the brick would not dry out and cause cracking. We used the term "dry from the inside out." Chopped grass or straw allowed this to occur.

Chopped seaweed provided plasticizing characteristics like the juice of the nopal and the drying characteristics of the chopped straw.

The Admixture Market

Pig urine was added to adobe exterior plaster in a village in Jalisco to prevent the adobe from washing off the wall in heavy rains. I have never found that it provided a benefit for stucco.

Molasses was commonly added to lime plasters to produce a harder surface, but as Portland cement started being added to plasters to decrease the setting time, the molasses was taken out, because molasses acts as a retarder for Portland cement hydration.

Modern Admixtures

I have several books on admixtures. One is 2.6″ thick. Since the publication of that tome in 1995, the information on admixtures has more than doubled. This chapter can only skim the surface of the knowledge that is available on admixtures. Prior to delving into admixtures, I will go ahead and state my bias. If you are using a good mix, admixtures are usually not needed. If you are using a bad mix, you should be changing your mix design. That having been said, I have used admixtures in the field, and I have incorporated them into mixes that are blended in a factory and shipped to the job in bags, bulk bags, and silos.

My father taught me that everything produced by man has a purpose, and sometimes that purpose is to sell. While there are many good admixtures that will improve a product, not all admixtures do what they are alleged to do. In 2007, I wrote a paper with my assistant, Pam Hall, entitled Variations in the Activity of Dry-Powder Water-Repellent Mortar Admixtures with Different Mortar Formulae. It was presented at *ASTM*'s 11th Symposium on Masonry. It discussed how some water-repellent admixtures met standards, but did not provide the service that the customers expected.

Water-repellent admixtures are often added to mortar to ensure that water does not pass through a brick wall. Most of the water that passes through a brick wall does not pass through the mortar, nor does it pass through the brick. It passes through the interface crack between the mortar and the brick.

To increase the water resistance of a brick wall, one should look for a way to increase the extent of bond between the mortar and the brick. Before you go out and spend money on the latest and greatest admixture, ask yourself whether it is needed and whether it will provide the service you expect it to provide.

ASTM C 926 (Standard Specification for Application of Portland Cement-Based Plaster) contains in the materials section Paragraph 4.6, which states "Admixtures—see 3.2.2." Section 3.2.2 states "admixture—A. material other than water, aggregate, or basic cementitious material added to the batch before or during job mixing." The only reference to admixtures in the reference documents is to *ASTM C 260 (Specification for Air-Entraining Admixtures for Concret*e). This tells me that literally any admixture can be added to Portland-cement-based stucco that the design professional specifies. There is much more information in the *Appendix* of *ASTM C 926*, but appendix material is non-mandatory.

ASTM C 270 (Standard Specification for Mortar for Unit Masonry) is a standard for mortar and not for stucco; but it is often used to specify stucco. It limits admixtures to classified admixtures [those meeting the requirements *ASTM C 1384* and *C 979* (pigments) and unclassified admixtures (everything else)]. When unclassified admixtures are used, the property portion of *ASTM C 270* must be met, and the admixture must be shown to be non-deleterious to the mortar, embedded metals, and the masonry units. The admixture market is wide open.

ASTM C 1384 recognizes the following classifications of admixtures:

Bond enhancers
Workability enhancers
Set accelerators
Set retarders
Water repellents

Pigments are recognized as an admixture and are covered by *ASTM C 979*.

Fibers are admixtures, but will be covered in another chapter.

ASTM C 270 developed the concept of unclassified admixture, since some popular admixtures do not fit into any of the above categories. Maybe I can legitimately market pig urine and lard as admixtures. I'll probably have to develop more creative names.

In some parts of the world, antifreeze compounds are added to mortars and stuccos to keep them from freezing during the initial set, but they reduce the ultimate bond and compressive strength of the stucco.

Bond Enhancers

When two layers of stucco are bonded together, the bond can be a combination of a chemical bond and a mechanical bond, or it can simply be a mechanical bond. It is commonly believed that if the base coat of stucco is over seven days of age, there will be little chemical bond between the two layers. To enhance the chemical bonding, *ASTM C 926* allows the second coat of stucco to be applied as soon as the first coat is rigid enough to accept the second coat without damage. Since there is no way to adequately test to see if the first coat is rigid enough, failures do not show up until much later and are often misdiagnosed.

Bond enhancers are often acrylic polymers that are added to the stucco mix to increase the ability of the stucco to bond to any surface. The formula for an EIFS base coat is one bag of Portland cement and one 5-gallon pail of acrylic or latex polymer. This material will literally stick to anything. Bond enhancement on a stucco job does not use that much acrylic polymer. Usually, if it is a liquid, somewhere between a quart and a gallon is used per bag of cement.

The same bond-enhancing admixtures can be applied directly to the stucco base coat, and serve as a glue layer. When used in this manner, they are not considered admixtures. If using an admixture in this manner, there are a couple things to keep in mind:

Use a product designed for exterior use. Products that re-emulsify are usually much cheaper, but they lose their effectiveness when the stucco is saturated after several days of rain.

Follow the manufacturer's instructions. Some require the second coat of stucco to be added within 24 hours of application. Some allow for extended time prior to application of the second coat of stucco. Using these materials is no excuse for sloppy workmanship.

If you are going to use a bond-enhancing admixture as glue (and many of them are used in this manner, no matter what the literature says) test it before you use it on the wall. Build a sample panel, slick down a portion of the base coat, allow it to cure for 7 days, and then apply the bond-enhancing admixture. Apply the second coat of stucco during the time frame listed in the instructions. After the panel has cured for seven days, cut the panel up and soak a section of the panel in water for a week. If you cannot separate the second coat from the first coat with the edge of your trowel, you have glue that does not re-emulsify. Boiling the sample for an hour will speed the test up, but will also produce a few failures that would not occur in the field.

Workability Enhancers

What is workability? How do you test workability? These were questions that came up during the development of *ASTM C 1384*. The basic conclusion was that the man with the trowel knows what workability is. We have laboratory tests, such as the flow test and the water retention test, that are related to board life and workability, but they don't actually test workability. There have been some fancy machines developed over the years to test for workability, but it seems like they've always come up short. It was hard for some of these inventors to admit that they could not build a machine that could tell us whether a stucco is workable, but the man with the trowel could tell us in 60 seconds.

Some years ago I had periodic dealings with a man who worked for a competing laboratory. While I never met him face-to-face, he complained that I was trying to steal his formulae for low-cost pozzolanic mortars. This was after we had two excellent low-cost pozzolanic mortars and an excellent low-cost pozzolanic stucco on the market. His management complained to my management, and I was instructed never to take another phone call from him. Anyway, he developed a low-cost masonry cement that he believed could also be used as stucco. He tested it, and it met all of the requirements for *ASTM C 91 (Standard Specification for Masonry Cement).* He quit his job, took all his notes and data with him, and filed a patent. In time he received a patent. When I tested the material, I couldn't get it to stay on the trowel. It met the specification, but it did not have any workability. I relate this incident because many of the manufacturers of workability enhancers do not have people in the lab with practical experience using a trowel. It is my contention that any PhD who is going to develop workability enhancers, should apprentice to a plastering contractor for several years. That's not likely to happen, so every workability enhancer should carry the label "*caveat emptor*." That's Latin for "let the buyer beware." When I joined *ASTM* many years ago, Bryant Mather, who was head of the Corps of Engineers Vicksburg Laboratory and was one of the great men in the concrete industry, asked me if I was one of the *caveat emptor* Nordmeyers. I admitted that I was. You will have to wait a few years until I write A Different Book of Stories to hear the story behind that comment.

Some workability enhancers are nothing more than air-entraining agents. During the period 1930 through 1970, many of the traditional masonry cements on the market were an inter-grind of Portland cement clinker, limestone (a filler), and Vinsol resin. The Vinsol resin was a milling aid (kept the fines from balling up on the milling media) and an air-entraining agent. Workability was achieved by the way the clinker and the limestone were milled together and by the addition of the air-entraining agent. I spent many years milling the components separately and then blending them and not being able to obtain the results I could achieve by milling them together. In the process, like Thomas A. Edison, I learned a thousand things that did not work and one thing that did work.

All that having been said, there are some excellent workability enhancers on the market. Most of them are not needed if a mixture of Portland cement and Type S dolomitic hydrated lime is used. If the mud is being produced from masonry cement or mortar cement, the workability enhancer should be checked for compatibility with those products. In many cases the workability enhancers lower the compressive strength of the final product. The mortar fats that are mentioned in a previous chapter fall into the category of workability enhancers. They are designed to be added to Portland cement to provide fluff and body. I know of several plasterers who are proud of their mixes that are so light that they seem to float onto the wall. They start with a Type S masonry cement that, when tested in the laboratory, has a compressive strength of 2,200 psi, and add a mortar fat that is designed to be added to Portland cement. The resulting mix has a compressive strength in the 1,000 psi range, and I can scratch it with a fingernail. Normally the stucco starts to degrade and crack if it is not protected with a good elastomeric paint within a few weeks of being installed. And we'll discuss it more in ***The Stucco Book—Forensics & Repairs***.

Set Accelerators

There are times when you want your stucco to set at a rapid rate. This may be when the weather is cold, and you want it to set before the wall can freeze and damage the setting stucco. It may be that you have limited time on a project. Some years ago I furnished a formula to a company which had a contract to stucco a house that was going to be built during a 24-hour period. The stucco had to be applied and cured so it could be finished after 6 hours without any efflorescence forming to mess up the finish. There are ways to do it, but these have tradeoffs in workability and cost.

The traditional set accelerator for cement products has been calcium chloride at dosage rates up to 2%. The action of the calcium chloride is to accelerate the hydration of the tricalcium-silicate within the Portland cement. It is cheap and it is effective. While the calcium ions and the chloride ions do not hurt the hydration of the cement, the chloride ions react with embedded metal including reinforcement and accelerate corrosion.

Over the past decade, several *ASTM* specifications have been adopted that limit the addition of chlorides to cement-based products. *ASTM C 1384 (Standard Specification for Admixtures for Masonry Mortars)* states in part:

"6. Chemical Composition

"6.2 At the maximum recommended dosage, the mortar admixture shall add not more than 65 ppm (0.0065%) water-soluble chloride, or 90 ppm (0.0090%) acid-soluble chloride to the mortar's overall chloride content ...."

Obviously, one cannot use 2% calcium chloride and meet this standard.

Using 2% calcium chloride in stucco can be compared to dipping galvanized stucco lath in sea water and then keeping it moist for a month. Deterioration of the lath will occur. Actually, it will be worse than dipping it into sea water, because in applying the stucco, the sanding action of the stucco as it is applied to the lath tends to grind some of the galvanizing off the steel lath. This exposes surfaces that are not protected by galvanizing to chloride ion attack.

Products used as set accelerators can function in one of two basic manners. They can increase the rate of the hydration process, or they can develop what is known as a false set. In the latter case, they have little or no effect on the hydration process, but will cause the stucco to appear to have set, allowing another coat to be added.

False set is usually caused by compounds that interfere with the sulfate balance within the cement paste. If an accelerator contains a significant amount of sulfate, it will give you a false set, but will not be increasing the rate of hydration, which is what you want from a set accelerator.

There are some accelerators that work, but produce a hazardous working environment. Sodium hydroxide and potassium hydroxide are the two main players in this category. Would you be willing to add Drano (sodium hydroxide) to your stucco and then apply it with a trowel and risk chemical burns?

Carbonates, such as sodium carbonates, also act as accelerators.

Soluble silicates, such as sodium silicates, act as accelerators. A whole chapter on the wonder drug sodium silicate is planned for ***The Stucco Book—Forensics & Repairs***.

Any time you add a sodium-based admixture, you can anticipate an increase in alkali efflorescence. This will wash off or can be brushed off, but when initially seen, it gives the impression that the final colors on your stucco job are fading.

Set Retarders

There are times when you need to slow down the set of stucco. Probably the most common time is when the temperature is 110^{0} outside, and there is a 10 mph breeze blowing that is drying the stucco out as you apply it with a trowel. You do not need a set retarder at this time; you need to be using hot weather stuccoing techniques and ensuring that you are keeping the stucco hydrated so it has time to cure.

One of the more common ways to retard stucco is to increase the hydrated lime and reduce the Portland cement. While this does not involve a set-retarding admixture, it is effective. The downside is that the ultimate compressive strength of the stucco is lower. For interior stuccoing, I have used one part white Portland cement and ten parts of hydrated lime. This gives me several hours of workability. I have also removed the Portland cement from the mix, kept the room humid, and kept the plaster on the wall workable for 24 hours. While I do not recommend this, I was pushing the limits to see what I could accomplish. As a research scientist, I was paid to find points of failure. That way I could warn the user about them.

If retardation is achieved by retempering the stucco, the ultimate strength will never be achieved, and you may end up with stress cracks in the stucco that are parallel with the direction of the trowel movement.

One of the most effective set retarders is sugar. It is so effective that many ready-mix trucks carry a bag of sugar with them to throw in the mixer drum in the event they get tied up in traffic and cannot dump their load. Also, go to a job where concrete finishing is done and ask the foreman whether he allows his workers to carry soft drinks out on the area being finished. It just takes a little.

Water Repellents

When working for Best Masonry and Tool Supply, I made several crude bowls from stucco. They were about the size of a butter tub. After three days of curing, I put water in them and placed them on the sales counter. I informed the sales staff that the bowls were made from our Scratch and Brown Stucco, and that they had been cured for 3 days in a moist environment, but nothing else had been done to them. Applicators got excited when they saw that the bowls held water and did not leak. This sold a lot of stucco for us, and the bowls had to be replaced regularly because they tended to grow legs and walk off. Periodically I would be asked to come to the sales counter and talk to a customer. The most common question I received was, "I use Alamo Type S for stucco. Will you make a bowl from Alamo Type S so we can see how well it will hold water?" My typical response was, "Talk to Alamo Cement Company and get them to make a bowl. I'm not in the business of promoting their products." Word would come back that they had talked to Alamo, and that Alamo would not make any bowls.

Our sales staff had copies of a paper I had written comparing the leakage through brick walls laid with our masonry cement, with a Portland/lime mix, and with Brand X masonry cement. We never revealed what Brand X was, but the mason who laid up the walls for us had used Brand X for years; and when he saw the results, he switched to our masonry cement. Additionally, he told everyone who would listen to him the identity of Brand X. In time, the regional sales manager for Brand X Cement Company came to the lab and asked to see the walls (by that time the walls were set up for show-and-tell for anyone who wanted to come and invest 15 minutes) and the data. I gladly obliged. His conclusion was, "You are exposing those walls to hurricane-like conditions. That is immaterial, 'cause we don't have hurricanes in San Antonio." He then incorporated that statement in his sales story. The effective panel area for each test was 12 ft^2. The data showed that our masonry cement and the Portland/lime mix leaked less than 10 ml of water over a 24-hour period. There are about 28 milliliters of water per fluid ounce, so there was literally no leakage through the walls. With Brand X, there were 24,600 milliliters of water over a 24-hour period. Our salesmen could point out that Brand X leaked about a thousand times more than our product. Needless to say, it was an extremely-effective marketing campaign.

Paul Harvey used to present "the rest of the story," so I guess I will also. According to modern-day specifications, brick veneer walls are built with a cavity behind the brick, with flashing, and with weep holes along the bottom row of brick. The walls are designed to handle that volume of leakage and discharge it to the outside. Don't do like my brother-in-law did and caulk the weep holes closed. All of the points of leakage we could identify were through the cracks at the interface between the brick and the mortar. In producing the stucco bowls, there were no cracks. Five years later, I made up a single bowl from Type S Brand X masonry cement. It did not leak, but the outside of the bowl did get damp after 24 hours.

Before you invest in a water repellent, consider how the water repellent will work with water flowing through a crack. Consider whether you need a water repellent in a stucco if it does not have any cracks.

Water repellents are of two basic types: pore fillers, and hydrophobic agents. The pore fillers often are like the stearates, which combine with calcium ions and form calcium stearate cement which plugs the pores. They are highly effective, as long as there are no cracks in the stucco. The hydrophobic agents are like wax. They cause water to bead up and not enter the pore space. They also are highly effective, as long as there are no cracks in the stucco. Many stuccos are highly effective, as long as there are no cracks in the stucco. After you have a stucco house and you find a crack in it, the contractor will tell you, "All stucco cracks." That means you have to figure out which water repellent will keep water from passing through a crack. Be sure to get it in writing.

At one point in my career, I wrote several papers concerning water repellents. I was criticized during the peer review of one of those papers because I had not included a highly-advertised dry-powder water repellent. Since I did not want to have to buy 5 gallons or 50 pounds of each water repellent, I had contacted the manufacturers and let them know what I was going to be doing and asked if they would donate a 1-pound sample for testing. Most of them did, but an attorney for the highly-advertised product required that I furnish him all test data before publication and that he have the final say in what got published. Because I did not want to do that, and because I was too cheap to go out and buy the highly-advertised product, it was not included in the study.

Another manufacturer of a water repellent will warrant that the resulting stucco or concrete is water repellent if a layer of not less than 4" in thickness is used. With concrete, this is a reasonable requirement, but you do not see 4"-thick stucco on a regular basis.

Stucco walls are usually built with some kind of backup. If lath is used, there is usually a WRB. Water that passes through the cracks in the stucco drains down between the stucco and the WRB, and drains out the weep screed. Most water problems with stucco come from inadequate flashing of windows, not from water passing through the stucco or passing through cracks in the stucco.

Pigments

If you want black stucco, the most effective pigment is carbon black. Before you use it, consider its double-first-cousin: graphite. Chemically, they are the same. Both are lubricants. If carbon black is used as a pigment, after the stucco has cured, and as it goes through its daily movement from temperature changes, a few of the carbon black particles will slip out of the stucco and blow away. With time, the black wall will fade to gray, and then to light gray. After a few years, you will not be able to tell that there is any pigment in the stucco.

The only other strong black pigment is manganese dioxide, but it gives only a gun-metal gray color to stucco. To get the black color that architects like, some pigment formulators blend manganese dioxide with carbon black. This gives a rich black color which after a few years fades to gun-metal gray. One pigment salesman told me that after the job is complete, no one ever looks at the colors anymore, so the fading is never noticed. When I was president of Rainbow Cement Company (the one incorporated in Texas, not the one incorporated in Georgia), we lost more than one sale because we would not use carbon black to produce a rich black color.

Pigments are usually used only in the finish coat for stucco. Better results are often obtained when the finish coats are purchased as a pre-blend. That having been said, there are people who want to mix their own finish coats and want to add their own pigments to the stucco.

Earth colors can be produced using iron oxide pigments. These are the most economical pigments to use. If a white stucco base is used, less pigment can be used, and a pastel color can be achieved.

Remember Nordmeyer's Law? The more dirt that is finer than 200-mesh that you throw into a mix, the lower the compressive strength will be. Pigment acts as dirt and will lower the compressive strength of the mix. Usually a dosage rate of between 2% and 8% is used. Below 2%, if the man mixing the mud makes a few changes, or if the man with the trowel finishes at a different stage, there will be a noticeable color shift. If you use a gray stucco base, the stucco base will be cheaper, but the colors on the wall will not be as bright.

To obtain the same color from each batch, weigh out the pigment. Measure the sand and the water. Add the pigment to the mix with the initial sand and water. Let it mix so each pigment particle is separate. Do everything the same way for each batch. Never re-temper. The ultimate color is determined by how many pigment particles can be observed on the finished wall. A clump of 100 particles gives the same color as a single particle, except that the clump of 100 particles may cause a streak as the trowel passes over it. A mason once told me about his secret to double or even triple the strength of a pigment: he added his pigment like I had told him to do it.

If one wants to get away from the earth tones and get to the blues and greens, he will need to move to chromium- and cobalt-based pigments. These are as stable as iron oxide pigments, but are usually much more expensive.

Within the past few years, several liquefied iron oxide pigments have been placed on the market. While I have not seen them on job sites yet, they will be an excellent addition, since they will provide for better dispersion of the pigments. Better dispersion means more consistent color and less pigment.

Some plasterers who use liquid paint pigments claim that they hold up in an acceptable manner. I've played around with those a little bit but have never run a test to determine whether they fade.

An interesting problem I had when using pigments with latex-modified cements was that the color was determined by the pigment-to-latex ratio rather than the pigment-to-paste ratio. When the amount of latex was changed slightly to change the consistency of the mix, the color of the stucco changed.

If you want the Old World look or the Tunisian look, mixing your own finish coat may be the way to go. If you want a more even pigmentation of a wall, you may want to learn how to mix and apply a fog coat. A fog coat consists of the fines from the color coat—the cement, hydrated lime, and the pigment, and is discussed later in this book.

Unclassified Admixtures

Finally, there are admixtures that do not fit into any category but they are being sold and being used. As long as they do not cause any harm, should they be allowed? *ASTM C 1384* does not mention them, but *ASTM C 270* does.

Antifreeze

If you are going to stucco in freezing weather, you do not want the mixing water to freeze within the first twenty-four hours after the stucco has been applied. To protect the stucco, some people have added various antifreeze compounds. This is a common practice in Russia, but it is frowned on in the US. Any substance which will dissolve into the mix water acts as an antifreeze, not just polyethylene glycol which is used as anti-freeze in vehicles.

Other comments

While special emphasis has been placed on not using chlorides in stucco, one should consider whether there are chlorides in polymers. Given time, many polymers will degrade; and if they contain chlorides, those chlorides can be released and can cause corrosion of the metal lath.

Any time you add an admixture to a bagged stucco mix, remember, you do not know the formula for that stucco mix, and there may be some component in that stucco mix that will cause a problem. One such problem will be discussed in ***The Stucco Book—Forensics & Repairs*** where a very common admixture was added to a bagged stucco mix, and it delayed the set of the stucco so we did not see the required 28-day strength until well past 56 days of age. Can you picture what happened during that 56 day period? Life is too short to have this kind of problem.

Practical Advice on Admixtures

When discussing admixtures with Nolan Scheid, he asked that I include a chapter with practical advice on admixtures. My theory on adding admixtures is if you have a good mix, you do not need most admixtures. If you do not have a good mix, you need to move to a good mix. That having been said, there are times that admixtures will help.

If you are mixing stucco on the job site and you need to accelerate the set, ask yourself whether you really need an accelerator, when doing so may inadvertently decrease the ultimate strength of the stucco. Usually you do not need one. If you need one, besides sodium silicate there are two ways that you can easily do it. Use Type III Portland cement (high early strength) or keep your materials warm and heat your mixing water to about 110° Fahrenheit.

If the weather is hot, you want to slow the set of the mix down. Usually the best way to retard the set of stucco is to work on the shady side of the building and start during the cool part of the day. With a few lights, there is no reason why the crew cannot be working at 3:00 am and be able to take off for the day in time to go to lunch. By keeping the sand pile shaded and moist and by using cool water or even ice water in the mix, the set time of the stucco is retarded.

If you use ice, be sure that all of it has melted before you dump the mud out of the mixer.

If you think you need a water-repellent admixture, it may be that you need to use better flashing and stuccoing techniques. Most leakage in a stucco wall is at improperly-flashed windows and other penetrations. The next most common place is through cracks. If you follow good procedure, you can flash penetrations correctly and eliminate that source of water. By following good lathing and stuccoing techniques, you will substantially reduce the number of cracks in the stucco. Do you need a water-repellent admixture? If you do, consider sodium meta-silicate.

If you add sodium meta-silicate to the mix water, it can increase the hardness of the stucco, retain water so the stucco cures better, and fill the pore space so the stucco is more resistant to the penetration of water.

If you have a batch of stucco that appears flat, you can often give it some body with a small squirt of liquid dish soap. This is an air-entraining agent. Be very careful, or I will write you up in the next version of this book. Remember Feret's Law. Air degrades strength as badly as water does.

If you need polymer-modified cement use a cheap exterior acrylic latex paint. Use from one quart to up to one gallon of paint. Dilute the paint with an equal volume of water. Place the diluted paint in the mixer with $1/3$ of the sand needed and then start adding the stucco concentrate. Add more water as needed to keep the mix workable. After all of the stucco concentrate has been added, add the remaining sand. Usually with polymer-modified cement, you use about two parts of sand for every part of cement.

Chapter 18
Fibers In Stucco

It used to be easy to decide what kind of fiber you needed to add to your stucco and the dosage rate. All you had to do was go down to the local tannery and ask how much ox hair they had. If it was priced right, you took it, decided how many batches of stucco you were going to mix, divided the ox hair into that many piles, and then added one pile of ox hair to each batch. Ox hair was the only fiber that was commonly used during the early days here in the United States.

When concrete needed to be reinforced, steel rebar and steel mesh were added. In hindsight, according to the "green" advocates, bamboo probably would've been a better choice than steel, but we had lots of cheap steel, and no one was pushing bamboo back then. But then, bamboo rots and you have to use a lot of it to get the same reinforcement you can get from steel.

Asbestos Fiber

Asbestos has been a common fiber since the time of Christ. While it is not mentioned in the Bible, more than one Roman statesman had a tablecloth made from asbestos. After a meal when everyone had wiped their greasy fingers on the tablecloth, the tablecloth would be thrown into the fire and then taken out and shaken.

Someone found that if asbestos, which is a micro-fiber, were added to concrete, greater flexural strength could be gained. By removing the large aggregate and carefully controlling the amount of asbestos, the asbestos siding industry was born. Asbestos siding could be thin, so it tended to be lightweight. It was tough. It could be textured. It could be colored. And it seemed to last forever. In some parts of the country most of the houses built in the late 1940s on through the 1950s had asbestos siding. Then researchers determined that asbestos can cause cancer. The brouhaha was about as bad as when researchers said that tobacco causes cancer. While some people maintained that asbestos had been used for millennia, and therefore was safe, the evidence that it causes cancer was overwhelming. The asbestos-cement industry died. Besides asbestos siding, that industry included asbestos roofing and asbestos pipe, to name just a few of the products. Meanwhile, there was a mad rush to find a fiber that was as effective and economical as asbestos fiber. So far, such a fiber has not been developed.

Fiberglass Fiber

The closest thing to be developed was fiberglass fibers, which were developed shortly before WW II. In this document we refer to the fibers as fiberglass fibers. We refer to products made from those fibers as fiberglass, as in fiberglass lath. It hit the concrete market like it was going to take over. The salesmen talked about how you could eliminate the steel mesh and just have the rebar in the concrete. This was going to cut down the amount of time that was needed to tie the steel together. On top of that, the fiberglass fiber was cheaper than the steel mesh. The salesmen also said that they were in the process of developing a system where the fiber could replace all steel and concrete. This was welcomed by those people whose job it was to pour concrete. But something happened, and the salesmen quit talking about replacing the steel rebar with fibers. Then they stopped talking about replacing all of the mesh with fibers.

Fiberglass fibers were used in structural stucco, which is commonly known as shotcrete. By using fibers that were 1/2″ long, it was possible to pump the mix and spray it onto a substrate. This technique is used for reinforcing highway embankments and mining tunnels. It was as strong as steel-reinforced concrete, but much easier to install, because it did not need forms, and it fit the existing contours of the substrate.

Glass Fiber Reinforced Concrete, as we currently know it, was developed from the structural stucco that was sprayed with shotcrete equipment. GFRC is used to construct panels, often curved decorative panels, for installation on the façades of buildings. These panels are stronger and much lighter than poured concrete of the same dimensions. So I guess those salesmen were right, they just didn't realize that they were not talking about concrete slabs.

With the successes, fiberglass fibers started to be recommended for use in stucco. The fibers did a good job of controlling plastic shrinkage cracking. A problem with the fiberglass fibers in stucco is they tend to be stiff and stick out of the wall like fuzz. Applicators soon learned that a torch would remove that fuzz.

There are different grades of fiberglass fiber. There is E-glass, which is cheaper. There is AR-glass, which stands for alkali-resistant. For a while, the fiberglass industry had the stucco fiber market to itself.

Polypropylene Fiber

Up until this time, the most common reason given for fiberglass fiber in stucco was to reduce plastic shrinkage cracks. Then fiberglass fiber got a competitor—polypropylene fiber. Soon there were allegations that the fibers increased both the compressive strength and the tensile strength.

Of course, the fiberglass salesmen were able to point out that polypropylene fiber tended to melt at a fairly low temperature; and if there was a fire, the polypropylene fiber would melt and not provide any reinforcement.

The polypropylene salesmen pointed out that when exposed to the alkali in fresh stucco mud, the fiberglass fibers get brittle after 24 hours; so, whether there is a fire or not, the fiberglass fibers are not there to provide any reinforcement to the wall.

Nylon Fiber

Nylon fibers have been around longer than the fiberglass fibers and the polypropylene fibers, but they had not been used in concrete. After polypropylene fiber was introduced to the concrete reinforcing market, it didn't take long for nylon fibers to be introduced to the stucco market. Both the fiberglass fiber salesmen and the polypropylene fiber salesmen ganged-up on the nylon fiber salesmen and pointed out that nylon fiber tended to be stiff, so the fibers stuck out from the wall. They also pointed out that nylon fibers could absorb up to about 6% moisture, so they would swell when the stucco was wet; and then after the stucco cured and dried, the fibers would shrink and any bond that developed between the fiber and the cement paste would be ripped apart. Whether this is true or not was immaterial, but the salesmen could make it sound very convincing.

Stealth Fiber

Then in the early 1990s the stealth fibers were developed. These were polypropylene fibers that were much finer than the monofilament polypropylene fibers we had been dealing with. They were measured in denier, which is a measurement of yarn fineness. Fibers that were 6- or 7-denier were virtually invisible in a stucco mix. The finer fiber produced more reinforcement per ounce of fiber used, because there was more of a tendency for these fibers to bend around particles of sand and because there was more fiber surface area exposed. They also lie very flat when troweled. We had some people tell us that these fibers were not even in the mix, until we showed them that there really were fibers in the mix.

I started a testing program of fiberglass, nylon, and polypropylene fibers of different diameters and different lengths. In this testing program, I used different concentrations of fibers in an attempt to find the ideal dosage for each fiber. The surprising thing I found with stealth fiber was that less than 2 oz /bag of stucco concentrate did a better job of preventing plastic shrinkage cracks than following the manufacturer's guidelines of 4 oz /bag of stucco concentrate. I talked to one of the technical people from the fiber company, and he told me that I had done much more work on the use of fibers in stucco than all of the fiber companies put together. He said there just wasn't enough money in it to pay for the research that was needed, so all of the stucco companies looked at their concrete test results and made educated guesses. The producer of Stealth Fibers changed their company name and changed the name they applied to the fibers. "Stealth fibers" has become a generic name for fine fiber.

Within the past few years, a 3-denier polypropylene fiber has come on the market. While I have not used it, I suspect it will be even better than the 6-denier polypropylene fiber that I have raved about.

PVA Fiber

PVA (polyvinyl acetate) fibers are another option. While they may have their place, I have never had success using them in stucco. First, to get the maximum benefit from them, they need to be pre-wetted. This causes a problem on a job site. If they are not pre-wetted, they tend to suck water from the mud after it is dumped from the mixer and cause early stiffening of the mud.

Next, consider the problem I had testing them in the lab. I needed to test at constant water/cement ratios, but was I supposed to include the pre-wetting moisture in that ratio or not? I could go on about the technical difficulties, but I will spare you.

Steel Fiber

When I first heard about steel fiber, I pictured something like steel wool that had been chopped up. I could not comprehend how it would not rust immediately, and then I was properly introduced to steel fiber. Each piece was about 1 1/4″ long by about 1/4″ wide and 1/16″ thick, with a narrow place in the center. When propelled from a stucco gun, it was a dangerous projectile. Such fibers, while they have their place in the shotcrete industry, do not belong around the traditional stucco industry.

Dog Hair

Back 10 years or so ago, my friend Kay, a dog groomer, decided to build a straw house. Since she had lots of dog hair available, she decided to use dog hair as the fiber in her stucco. It worked real well, except that after a couple humid days, one can sometimes smell the slight aroma of wet dog. Since all of her stucco with dog hair in it that can be exposed to high humidity is on the exterior of her home, it did not cause a problem.

Fiber Background

Polypropylene and nylon fibers initially came as monofilament fibers, similar to chopped fishing line. The general dosage recommendation for all of the fibers used to produce stucco was about 4 oz/80 lb bag of stucco concentrate.

Since the polypropylene fibers did not bond to the cement paste the way the fiberglass fibers did, they did not provide as much reinforcement. To increase the bond, the polypropylene fibers were distressed (roughened on the outside) so there could be a better mechanical bond. These distressed fibers were referred to as fibrillated fibers. They worked wonderfully in concrete, but in stucco they had a tendency to ball up in the mixer to form what we called bird nests. Even a small bird nest in the stucco would seriously mess up the wall.

Meanwhile, the fiberglass fiber bundles did not break down readily into individual fibers without larger aggregate, so there were fiber bundles that showed up on the wall. When you can see fiber bundles 10 feet from the wall, the customer is not particularly happy.

Let's talk a little bit about those fiberglass fibers that get brittle when exposed to the water found in the stucco mud. If I wanted to impress you, I would use the correct term of interstitial fluid, but I am not trying to impress you. Fiberglass is non-crystalline glass, so it is slightly soluble in an alkaline solution. As it dissolves, it reacts with calcium ions that are liberated when Portland cement hydrates. This forms hydrated calcium silicates, which tend to clog the pores in the stucco and make it resist the penetration of water. If the fiberglass fibers are broken down into individual filaments, and if the alkali concentration in the stucco is high, the fibers can get brittle. As a result, they can become useless after the first 24 hours. If, however, fiberglass lath is used, and if low-alkali stucco is used, you end up getting some pozzolanic bonding between the fiberglass lath and the stucco. A plasterer told me that he firmly believes that when he uses fiberglass lath, his stucco become six times as hard. I have not run any tests on this to see whether this is the case; so when you try it out, see what you find and let me know.

How long should fibers be? For stucco I have found that 1/2″ to 3/4″ fibers work best. If the fibers are longer than 3/4″, there is more of a tendency for bird nests to form in the mixer. If the fibers are shorter than 1/2″, we don't get quite as much bonding as we would like.

It is my contention that the only reason for using fibers in a conventional stucco mix is to limit plastic shrinkage cracking. By providing a mat of fibers in the stucco, the stresses are spread out, and hairline cracks do not form during the first 24 hours after the stucco is placed on the wall. After the stucco gains more strength and higher stresses are placed on the stucco, it tends to resist cracking. If the fibers are not used, during that first critical 24 hours, microscopic hairline cracks may form. While they are not noticeable, at a later date when more stress is placed on the wall they may expand and become noticeable. Fibers are like many admixtures. If a good formula is used and if good procedures are followed, fibers often are not needed. When was the lat time you observed good procedure being followed?

Chapter 19
Applying Stucco With a Trowel

Prior to mixing stucco, it is appropriate to assemble all supplies, tools, and people that you may need. I have been to more than one job that was delayed because there was not enough gasoline to mix the first batch. I have been to others that shut down in the middle of the morning because the mixer ran out of gasoline. I have been to other jobs where plasterers had not shown up by the time the mud was mixed. If you need a list to ensure you have everything you need, make one. Since this chapter is about applying stucco, we will now list some of the items that are needed to mix the stucco.

Stucco Tools

For applying stucco, the minimum requirements are
a hawk,
a trowel,
a wheelbarrow,
a concrete hoe
a shovel,
a mortar board,
a scratch tool,
a water source, and
personal protection.

A darby and a browning rod make the job go a lot easier. Most people should start with a trowel that is four inches wide and about ten inches long. A swimming pool trowel is oval rather than rectangular and is usually easier to learn with. As one gets used to using it, it is appropriate to graduate to a larger trowel; and as one learns to not drag the edge of the trowel in the mud, it is appropriate to switch to a rectangular trowel. Both changes will allow mud to be applied to the wall in a more efficient manner. A new plasterer should probably start with a twelve-inch-square hawk and with time graduate to a larger one.

A darby is a three-foot-or-longer device with a handle on each end that is used for leveling the stucco. Often it is used with one end riding on a control joint of casing bead, so the stucco can be leveled to be in a plane with the outer edges of the control joints and casing beads. A browning rod provides the same function as a darby, but it is usually longer and does not have handles. Now, we know why a scratch coat is called a scratch coat, but why is a brown coat called a brown coat? The answer is easy; it is leveled with a browning rod. Do you know why a browning rod is called a browning rod? I have no idea.

As far as personal protection is concerned, you should wear safety glasses and waterproof gloves. An apron keeps the stucco off your pant legs. Remember, stucco is alkaline and can cause chemical burns. These burns usually are not felt until after they are well advanced.

A mixer makes mixing stucco much easier and faster. If you have rented a mixer, you will also need:

Gasoline or electrical power for the mixer,
Oil if the mixer uses a gasoline engine,
At least four 5-gallon buckets, and
An additional wheelbarrow.

On large jobs it is common to apply the stucco via a spray rig. There are a number of systems on the market for spraying stucco, ranging from the sprayers sold by Mortar Sprayer (http://www.mortarsprayer.com/), to pumps that pump stucco through a three inch hose to a spray gun.

With Mortar Sprayer's system, the gun is attached to an air line, and it is filled by dipping the gun into a wheelbarrow of stucco, or it can be filled by a person shoveling mud into the hopper. The trigger is then depressed, and the mud is blown onto the wall. While working on the Tomko House (a volunteer project on which over 200 of us finished a stucco mansion after the owner/builder fell from the roof and died when the house was 10% complete), a mortar sprayer was used to apply stucco. On one occasion a young lady was standing on a ladder spraying stucco and doing a good job. One of the young men, who had no experience but knew more than all of the rest of us put together, came up, and I assume he was attempting to flirt with her as he started pointing out places she had missed. She leaned over to see better, asked, "Where?" and inadvertently triggered the mortar sprayer. The young man was plastered from the top of his bald (or shaved) head down to his upper thighs. Some of us cheered. The young lady was so embarrassed she never came back. The young man did not learn anything.

On the upper end of the spraying systems are trailer-mounted combination mixers, pumps, air compressors, and guns that can deliver up to 12 yd3 of stucco per hour. Before you go out and buy such a system, remember that there is a learning curve associated with them. At one point I owned a relatively-small stucco pump. When I purchased it, I was informed that the original owner was a plastering contractor who figured he could spray a house in a day, rather than spending a week hand-plastering it. He found that he could, but the clean-up took him about a week, rather than a few hours. He sold the system and went back to hand-plastering.

If one uses one of the big pumps and a 3″ diameter hose, it is common for the hose to be 200′ long for a residential job. Such a hose will hold 10 ft^3 of mud. That means that one must prime the system with 10 ft^3 of mud; and if there is a breakdown one must clean 10 ft^3 of mud out of the hose. If that mud is not removed, it will set up in the hose. If there is even a small leak in a hose, the stucco will thicken and can plug the hose. If the stucco is not adequately mixed and stiffens slightly after leaving the mixer, it can plug the hose. Then there is the problem of moving the hose as the job progresses. The hose will hold about 8 pounds of stucco per foot of hose. The man with the gun cannot move hose and control the gun, so a second person is needed just to move the end of the hose.

Between the Mortar Sprayer and the big rigs is a beautifully designed system developed by BlastCrete Equipment Company called the MasonMate.

(http://blastcrete.com/products/MasonMate/Mixer-Pump-for-High-Rise-Block-Fill/)

It is designed to spray stucco and to fill concrete masonry units with grout. It is small enough to be hauled in the back of a pickup truck, and comfortably fits into a freight elevator. It can be set close to where the stucco is being sprayed, and often only 25′ of 1 1/2″ hose are needed. It is designed to be fed with 1,000 pound bulk bags of sanded stucco mix. It can operate in a room with an 8-foot ceiling. It takes only 1/3 ft^3 of mud to fill that hose. If a problem occurs on a job site, only 1/3 ft^3 of stucco needs to be cleaned out of the hose.

The Scratch Coat

When applying stucco, plan on applying stucco to an entire panel before you stop. This reduces the cracking of the stucco and also produces a more consistent finish coat. Some plasterers like to start at the top of a panel and work down; others like to start at the bottom and work up. While most can give reasons, it tends to be a regional preference. For a finish coat, starting at the top and working down reduces the chance for drips on a soft finish coat, but more skill is needed to select the load on the trowel to end up the sweep without having extra mud. When starting at the top and working down, the trowel is never moved in a downward stroke. Rather, the trowel is loaded and moved in an upward stroke, then the next stroke starts lower on the wall and moves upward. If the trowel is moved in a downward stroke, the mud has a greater tendency to fall off. When applying the scratch coat, keep the leading edge of the trowel far enough away from the lath so the trowel does not catch on the lath laps. To prevent this problem, some plasterers lap the lath so the upper sheet of lath is stuck behind the lower sheet of lath. This works fine, but be sure that the paper, if paper-backed lath is used, is lapped to shed water to the outside.

Apply the scratch coat thickly enough to fully embed the lath, and apply enough pressure so the mud is pressed against the WRB. If you do not have enough mud on your trowel, you may think you have a good coating of mud, but the trowel is riding on the stucco lath, and you are only filling the holes in the stucco lath. The scratch coat should extend about one-eighth-inch beyond the lath and you should not see the texture of the lath after the mud has been applied. If you see the texture of the lath, either the mud is too thin, or the mud is too fluid and is sagging.

After the scratch coat has started to set, a scratch tool should be used to place horizontal scratches into the surface of the stucco to increase the mechanical bond when the next coat of mud is applied. Besides increasing the mechanical bond, the scratch tool opens up the slicked portions of the mud and allows the stucco to breathe and to lose some moisture, and it removes the high fines surface coating that develops when stucco is over-troweled.

If you are applying mud to a surface without lath, apply enough pressure so there is a good bond between the mud and the substrate.

The Brown Coat

We now reach a point of controversy. I prefer to allow the scratch coat to cure for at least forty-eight hours before adding the brown coat. *ASTM* and other standards now allow the brown coat to be applied as soon as the scratch coat is solidified enough to accept the scratch coat without damage. That is where I have a problem. Who makes that judgment call? If the sheathing is stiff and has no give, and if the scratch coat is applied so that the mud fully embeds the lath and there are no voids between the back of the stucco and the WRB, then the scratch coat is not damaged with the addition of the brown coat.

If those criteria are not met, then applying the brown coat can crack the scratch coat, and the applicator never knows that the scratch coat has been damaged.

Another and probably more serious problem also occurs. When stucco cures and dries, it shrinks. In the process, cracks (called plastic shrinkage cracks) can occur in the scratch coat. These are the cracks that look like the cracks in the bottom of a mud puddle that has dried up. They occur within the first forty-eight hours after the mud is applied. They usually occur within the first twenty-four hours. If plastic shrinkage cracks occur in the scratch coat after the brown coat has been applied, there are no forces on the job site that can keep the brown coat from cracking. This modification in the standards was instituted because of value engineering. Value engineering is finding ways to save money, even if the result is a lower-quality product. If you were a homeowner, how much would you want to save before you would be thrilled with a home with plastic shrinkage cracks in the stucco? If the applicator will let the scratch coat cure for forty-eight hours before applying the brown coat, most of the problem will be eliminated. At the same time, half of the potential efflorescence will be eliminated.

Apply the brown coat in a manner similar to the scratch coat, and bring this total stucco coat out to where it is even with the outer edge of the casing bead. Use the darby or browning rod to level out the surface of the stucco. Use the outer edges of the casing bead, control joints, and weep screed as screed points. You want a flat surface, but you do not want the surface slicked down. If it is slicked down, use the edge of the trowel to back-drag across the stucco to end up with a sandpaper-like surface.

Chapter 20
Spraying Stucco

Stucco may be applied with a trowel, but it may also be applied with a sprayer. A sprayer may be as simple as a hand-held Tirolessa USA Stucco Sprayer or as complex as a double-piston pump. The method of getting the stucco to the gun differs with each system, but once at the gun, each uses air pressure to propel the stucco onto the wall. In this chapter we will briefly discuss the different transport systems and then discuss the spraying of the stucco onto the wall.

Delivery Systems

There are three basic delivery systems—the dry system, the wet system, and the wheelbarrow system.

The Dry System

The dry system, often is often erroneously called the Gunite system, uses low pressure and a high volume of air to transport a mix of cements, aggregate, and admixtures from the mixer to the gun. Water is added at the nozzle. An advantage of this system is that the water/cement ratio can be controlled. By keeping the water/cement ratio low, this allows a thicker layer of stucco to be built up and results in stucco with a higher compressive strength.

Often the dry system is used for structural stucco, such as stabilizing mine tunnels, building rockscapes and swimming pools, and stabilizing highway embankments.

The dry system hoses usually do not develop leaks until they wear out, since the air moving through them is low pressure. When a leak does occur, there is no doubt about it because of the dust cloud that forms.

Depending on how the nozzle man adjusts the water addition, the area around the nozzle can be dusty. Wear a dust mask. Protect your eyes. Since the nozzle man has full control over the amount of water used, and thus the water/cement ratio of the product, he can apply it so dry that there is not enough moisture for the cement molecules to properly hydrate, and he can apply it so moist the mud flows off the wall before it has an opportunity to set. Usually he applies it somewhere in between, but there are still problems that can develop. When water is added beyond what is needed for hydration (a water/cement ratio of about 0.35), the ultimate compressive strength of the product is reduced. The amount of water used needs to be based on the required compressive strength of the product and the workability that is needed to finish the surface.

Rebound happens. It can be picked up and recycled.

The Wet System

The wet system uses a pump to move mixed plastic stucco through a hose. Shotcrete, also known as structural stucco, is a subset of this system. The plastic stucco can be moved by pressurizing the hopper holding the stucco to force it through the hose, by a rotor stator pump, by using a peristaltic pump (squeeze tube pump), or by several varieties of concrete pumps.

As far as I know, the last pressurized hopper pump was one that I built 15 years ago and discarded 6 months later after building a peristaltic pump. The pump had two pressure containers and when depressurized could be filled with stucco. After closing the lid and being careful not to get mud on the threads, the container was pressurized. The amount of pressure used was related to the length of the hose. Since the unit was small and maneuverable, I usually used 25′ of hose and a pressure of less than 40 psi. When the man on the wall handling the nozzle (me) was ready for stucco, the pump operator turned a valve and stucco flowed through the hose. Meanwhile, the pump operator opened the other container, filled it, closed it, and pressurized it. If the pump operator did not switch to the other tank when the first tank was empty, the rate the stucco moved through the hose increased. As the hose emptied, the sound the system emitted increased in pitch. If the man with the nozzle (that was always me) did not brace himself, he could be knocked off the scaffolding. If the mixing crew and the pump operator would keep working, rather than slowing up and placing bets as to whether I would be knocked off the scaffold by the hose emptying, I could apply 1 ft^3 of stucco per minute. That equated to 2 1/4 yd^3 per hour and 24 ft^2 of 1/2″ thick stucco on the wall per minute. Can you understand why I switched to a peristaltic pump and why I have never attempted to market plans for building a simple pressurized hopper pump?

A rotor stator pump usually consists of an auger-type shaft which is fitted with a heavy rubber-type sleeve. As the shaft rotates, the mud is carried through the sleeve and forced into the hose on the other side. These were some of the most common early wet system pumps but are no longer as common as they once were. A problem with them is that if larger diameter aggregate gets into the pump, it can damage the rubber sleeve. If the pump is not cleaned out after each use, starting the pump can also damage the stator. Once upon a time, I ended up with such a pump. The previous owner told me the lengths he always went to in order to clean the machine after using it, so I did not check it out before buying it. By the time I got it home and checked it out, I discovered that the replacement parts for it were no longer on the market. When discussing the condition of the pump with the previous owner, he told me that after he had used it the last time, he had loaned it to a friend.

On another occasion I built a system using the largest rotor stator pump Granger sold. It was limited to a particle size of 3/16″ and worked wonderfully for spraying top coats on stucco. It was easy to increase the rate of pumping; all I had to do was change pulley size. I also learned that speeding up the pump reduced the life of the stator to about 3 hours. I would still be using that pump at its slower speeds, but about 10 years ago I loaned it to a friend who has failed to return it.

The peristaltic pump uses rollers on a flexible hose to push small amounts of stucco through the hose. If the peristaltic hose is ruined, it is easily replaced. If it is belt driven, usually the belts slip if there is a clog in the system. As a result, belt drive peristaltic pumps are no longer on the market. Hydraulic motor drives allow a smooth ramp-up in speed and allow for the pump to be reversed to lower line pressure if a clog occurs.

In my opinion, the most innovative peristaltic pump on the market today is Blastcrete's MasonMate. It was designed to mix and pump grout, but also pumps stucco. Rather than a mixer that takes up a lot of room, it uses a high sheer mixer. It can fit in the back of a pickup, or in a freight elevator. It can be fed with 1,000-pound bulk bags in an area where the ceiling height is eight feet. Since it is so maneuverable, a 25-foot-long 2″ diameter hose is all that is needed in most areas. The pump delivers up to 400 psi of pressure, so if one needs to pump up several floors, more hose is added and the pump can handle it. Twenty-five feet of hose holds only about $1/2$ ft^3 of stucco and as a result weighs about 75 pounds. With a variable speed drive, it can easily deliver from 0 to 6 yd^3 of stucco per hour. That reduces the number of people that are needed to move the hose, so the nozzle man can do his job, and it is simple to clean the hose out if the job is shut down unexpectedly. The only downside is that the stucco mix needs to be delivered to the job in a sanded form; and with all of the positives, that is a minor downside.

The work horses of stucco pumping are the concrete pumps. There are single-cylinder pumps, double-cylinder pumps, something called a swinging ball valve pump, and a few other configurations. If a single-cylinder is used, it pumps a stream of stucco, is idle while the cylinder refills, and then pumps another stream of stucco. These pumps can generate considerable pressure and should be used only by people who are trained in their use. I've watched these pumps deliver stucco 300′ laterally and 7 stories vertically without grunting. I have also seen the pumps fail, leaving 300′ of 4″ hose full of stucco.

That is about a cubic yard of stucco, which weighs about 3,600 pounds, and is difficult to remove before the stucco sets up and ruins the hose.

A problem with pumping stucco through hoses is that if the hose leaks, even a little bit, liquid water will exit, and the stucco will get thicker and harder to pump. Harder to pump means more pressure in the hose and more water leaking from the system. Another problem is when stucco is used that stiffens shortly after it is mixed. Most pump operators are very particular with the stucco formula they use.

The wet system is not as dusty as the dry system, but one should wear a dust mask and eye protection.

The Wheelbarrow System

The third system uses a wheelbarrow to transport the stucco to the gun, and the gun is dipped into the wheelbarrow to fill it, or a shovel is used to fill the hopper for the gun. Since there are no stucco delivery hoses, there are no heavy hoses to drag around. If the job is shut down, there are no hoses to clean out before the stucco sets.

Www.MortarSprayer.com has a hopper gun on steroids. It comes with about a 3-foot handle and can deliver stucco to a wall about as fast as it can be loaded. These sprayers are priced right, and the learning curve is very short, so if you have house-sized jobs and would like to get into spraying stucco, this is an ideal way. These sprayers have a distinct advantage over the wet system. The mix does not need to be pumped through a hose, so a harsher mix can be sprayed than can be sprayed with the wet system.

As with other spray systems, one should always wear eye protection.

Applying Stucco

Some years ago I was watching a crew apply stucco to a Structural Concrete Insulated Panel (SCIP) wall that was about

22 feet high at the gable and about four feet lower on the side walls. Stucco was being applied by a crew that had been described to me by a college professor as an expert crew. Normally when applying stucco to SCIP, it is appropriate to apply stucco in critical areas to stiffen the wall before applying it to the full height of the building. I prefer to see it applied to encapsulate the mesh, and then after that has cured, to bring the stucco out to the desired thickness.

They had a scissor-lift. Rather than spray the corners first, and then spray horizontally to stabilize the wall, they started at the bottom of the wall and sprayed vertically. Not only that, for some strange reason, their first pass was the 22′-high section of wall. They were applying about 1.8 inches of stucco monolithically to the wall, rather than splitting it into two coats. The panels had a 2″ mesh that was located about 3/4″ from the EPS foam base. Three quarter inch PVC pipe (exterior diameter of 1.07″) were attached to the mesh and spaced vertically about 3′ apart as screed points. The concept was that after the scissor-lift had moved along the wall, a person with a darby would screed the panel to eliminate any high points. There was only one scissor-lift, so I wondered what the man with the darby would be standing on. When the sprayer reached about 10′ from the floor on the first panel, I noticed some telegraphing of the mesh through the stucco. This is a sign that the wet stucco is sagging. When the sprayer reached about 20′ from the floor, the stucco started to move. At first it moved slowly, and then it increased in speed and volume until it hit the concrete floor and headed horizontally towards the scissor-lift. I thought the scissor-lift was going to turn over, and apparently the man on it thought the same thing. Anyway, it did not. The next day the general contractor was looking for another stucco crew.

When spraying stucco, always remember you can apply much more mud with a spray system than you can by hand. If something goes wrong, it usually goes wrong on a much larger scale than it goes wrong when trowel-applying stucco.

A residential stucco contractor in central Texas bought a rotor-stator pump some years ago. It took him a week to plaster a house, and the salesman convinced him (he wanted to be convinced) that he could do the same amount of work in a day with a stucco pump. With the first house, he was able to spray it in a day, but it took nearly two weeks to clean up the mess. When I spent some time with this contractor about five years later, all of his mud was going on by hand, and he did not mention using a gun.

At a job site in the California high desert, I observed a man with a water hose cleaning the soffits within 15 minutes of the wall being sprayed. The crew leader told me that the developer did not want any stucco on the soffits. I wondered if the developer objected to gullies running through the stucco that were created as water flowed down the wall.

These three incidents illustrate that just because you have a stucco pump and stucco, you are not ready to apply the stucco. All windows and other things that should be protected need to be masked. The masking that is normally done with applying mud with a trowel is not even close to adequate. You need serious protection. If you can blast the protection off with the full blast of a water hose, it is not sufficiently attached. If something is within 3′ of the nozzle, it will get sprayed. Using shields to protect areas works in theory, but from a practical standpoint, it leaves a lot to be desired.

Prior to starting to spray, it is necessary to have an adequate crew. One man needs to be at the mixer/pump. Except for the MasonMate and smaller systems, a second person needs to back up the nozzle man by moving the delivery hose as necessary. Of course, someone needs to be operating the nozzle, and someone needs to come behind to darby or knock down high points in the stucco. For your first several jobs, it does not hurt to have a few more people around to keep things moving.

Mud is delivered to the gun at relatively low pressure. In most cases it will just flow out the nozzle; and within a few inches, gravity takes over and pulls it down. Air is added at the nozzle to propel the mud from the nozzle to the wall and to spread the mud into a cone-shaped pattern.

There is an art to controlling the nozzle. It needs to be kept moving at all times, and it needs to build up evenly over the face being sprayed. If the air pressure at the nozzle is too high, there will be excessive rebound of the mud. If the air pressure is not high enough, the cone-shaped pattern will not develop. If the nozzle is aimed directly at the lath, the amount of rebound increases. This rebound comes directly back towards the nozzle man and, as a result, the nozzle man's safety glasses are soon coated with mud.

Stand back. If you hold the tip of the nozzle within 6 inches of the wall, the amount of rebound increases dramatically, and it becomes close to impossible to apply an even coat. Initially, keep the nozzle twelve to fifteen inches from the wall. With time you can adjust this distance based on the flow from the nozzle and the air pressure being used. Ideally the mud should hit the target with enough force to stick in place and densify, but without enough force to rebound.

Now for a few numbers to amaze you. If you are spraying at a moderate 1 ft^3 per minute, that rounds off to needing 22 bags of concentrate mixed with 66 ft^3 of loose, damp sand each hour. If each batch is going to mix for five minutes, a two-bag mixer will not keep up. If you mix for less than five minutes per batch, you run the risk of plugging the pump line. If the cement and sand are carefully staged, you will want two men mixing mud; otherwise, your crew will be stumbling over each other.

Spraying at that moderate rate and applying a coat that is one inch thick, 24 ft^2 of wall will be covered every minute; but you will get better results if you apply 1/2″ per coat, so you will be covering 48 ft^2 per minute. That is six linear feet of a wall that is 8′ tall. If the cone of spray is 12″ in diameter, it covers about 0.8 ft^2. With overlapping to obtain an even coating, you will probably need to be moving the nozzle at about 2 mph. This is doable, with a trained crew, but since greed is the primary motivating emotion of many of us, you might be tempted to pay the extra money and purchase a pump that will move 6 ft^3 of stucco per minute. That is just over 12 yd^3 per hour. Bigger is not better if you are stuccoing residential construction with all of its short walls and penetrations in those walls. Yes, the larger pumps can be slowed down, but....

Approximately fifteen minutes after an area has been sprayed, pass a darby up the wall while it is anchored on screed points. This will knock down the highpoints. If you delay, you may have problems knocking down the high points, and then you will have to add a thicker coat of mud than you planned. Some plasterers leave the wall like that; others trowel the wall to densify the stucco. Densifying the stucco reduces plastic shrinkage cracking; but if the troweling is delayed until the initial set has started, it stresses newly-formed bonds between the cement molecules. If that occurs, cracks can show up the next day that are perpendicular to the path of the trowel.

Leveling and troweling stucco is an art, not a science. As a scientist, I would love to specify that 15.35 minutes after spraying a wall, it should be darbied; and then exactly 47.73 minutes later, it should be troweled. If I did this, I would be laughed off the job site. Seriously, one of the reasons for building the practice panels is to gain an understanding of when different steps need to be done. Running the darby over the wall to level the stucco can be done as soon as the man with the sprayers moves out of the way and the delivery hose and air lines are not in the way to trip over. By delaying about fifteen minutes, if the wall is going to absorb any moisture out of the mud, the process is substantially on its way. Usually if the wall is going to absorb moisture out of the mud, the mud should be applied a little juicy.

When troweling the darbied surface, if the trowel does not move freely over the stucco and level any minor surface imperfections, you have waited too long. If you start the process too soon, it is easy to slick the wall down so a top coat cannot mechanically bond to it. Temperature, humidity, wind speed, wind direction, and sunlight are all environmental factors that will impact the optimum time for doing these jobs.

Chapter 21
Plant-Ons and Other Decorations

A plant-on is an addition to a stucco wall which sticks out from the wall and is added after the brown coat has been installed. While the final appearance may be the same as an addition that is formed with lath, the method of achieving it is considerably different.

Wood Plant-Ons

One of the easiest ways to form a plant-on is to attach a piece of lumber to the wall, and then paint it a contrasting color. A plant-on may be made of wood and then stuccoed with finish coat stucco. During the first 6 months, it usually looks fairly good, but after that, the movement of the wood results in the stucco delaminating from the wood and cracking.

If one is going to use wood and stucco over it, one should attach a piece of fiberglass cloth to the brown coat on both sides of the piece of wood. Before you do that, consider the potential problems and decide that EPS Plant-Ons are a much better option. Some years ago I was in New Orleans looking at an apartment complex where the plant-ons were cypress 2×12s. To do a good job, kiln-dried cypress was used. To make a good job even better, each cypress board was attached to each stud it crossed over with two nails that were spaced about ten inches apart. As the cypress was exposed to New Orleans humidity, the boards expanded. Rather than being 11 1/4″ wide as they were when they were installed, they were over 11 3/4″ wide. They literally were tearing themselves off the wall.

On another job, they used green cypress so they would not have the expansion problem, and they solidly attached the cypress. When the sun hit the south side of the building after a few low-humidity days, the cypress cracked. If it were not for black widow spiders, I could have put my fingers into some of the cracks. One of the engineers suggested using lag bolts to pull the boards back together, then he stopped and said, "Forget I ever mentioned that." There is no power on a construction site that can prevent wood from moving when the humidity changes.

If you are going to use wood as plant-ons, consider doing it in a proper manner. After the WRB is installed, install treated lumber nailers wherever you want a wood plant-on. Usually you should use one inch thick lumber, since most stucco is installed approximately 3/4″ to 7/8″ thick. The treated lumber nailers should be at least 2″ narrower than the wood plant-ons. Stucco the wall. If the wood plant-ons are to be horizontal, their upper edges should be beveled to assist in shedding water. Paint the wood plant-on with at least two coats of a good grade of exterior paint. After the plant-ons are dry, apply a good grade of sealant to the treated wood nailer and to the stucco on each side of the treated wood nailer. Place the wood plant-on and nail or, better yet, screw it in place. Come back and touch-up the paint on the plant-ons.

EPS Plant-Ons

When EPS (expanded polystyrene) came on the market some years ago, it did not take plasterers long to learn that EPS does not move like wood. It is also much lighter, and it can be formed into many more shapes. To use EPS as a plant-on, the EPS is glued into place and covered with fiberglass mesh. Polymer-modified cement (think EIFS basecoat) is then used to coat the fiberglass mesh and bond it to the brown coat and to the EPS. The fiberglass mesh needs to extend out from the EPS six inches to give adequate bonding. A cottage industry has developed to produce EPS plant-ons with the fiberglass mesh attached to them. You can select from hundreds of shapes, or you can have them custom designed and shaped.

Anywhere there is a joint in the plant-ons, the joint needs to be covered with fiberglass mesh that extends six inches either way from the joint; otherwise, a crack will develop.

To successfully use EPS plant-ons, there are a few things to consider. If they are more than 8′ off the ground and extend out over 3 inches, they may provide a place for pigeons to roost and nest. If they do not slope away from the building, they may collect water and feed it into cracks in the stucco. If they are continuous over a control joint or an expansion joint, they will probably crack or be torn loose from the stucco.

After the EPS plant-on is secured to the wall, the wall is finished with texturing and coloring coats (may be combined).

Reveals

Reveals are indentations into the stucco. A common reveal is a horizontal groove. It is recommended that when reveals are installed, there be a break in the lath and that the lath lap over on the top of each of the flanges. It provides a sharp-looking groove. Usually reveals are made from aluminum or PVC, and may be made in one piece or in two pieces. A simple reveal can be made with two pieces of casing bead placed parallel to each other and the desired distance apart.

If you have an inexperienced man with a trowel in his hand, warn him to not fill in the groove in the reveal. It is a pain to remove stucco after it has set.

Other Decoration

Especially in custom homes, sometimes the home owner will want his children to place their handprints in the stucco. This gives a permanent record of the sizes of their hands when the wall was stuccoed. If a deep handprint is wanted, it needs to be in the brown coat. In finishing the wall, care needs to be taken that the handprint is not covered up. The man with the trowel in his hand can leave the texture coat off near the handprints. If the wall is to be finished with an acrylic texture coat, the area of the handprint can be finished with an acrylic smooth coat of the same color. That is a euphemism for acrylic paint.

Along the same line, any number of decorations can be added to the wall, such as footprints or leaves. To put the imprint of a leaf into a wall, with the stucco wet, place the leaf on the wet stucco and run the trowel over it to embed the leaf into the surface of the stucco. STOP!! If you continue to trowel over the area, you will bring fines to the surface and cover up the leaf. Get out and play with leaves and stucco and develop the technique you want to use. Since the leaves are very thin, you will want to embed them in the outermost layer of stucco that you are adding.

You can embed branches, wooden signs with letters routed into the wood, sea shells, arrowheads, or just about anything else you can think of. These embedded objects can be scattered over a wall, or they can be used to outline a window or door. How about a brass plate beside the front door that instructs visitors to take their shoes off before entering?

Some years ago a friend who was building a straw bale home collected cow chips. She alleged that she was going to plant them on the stucco and then paint them a gold color. She was inspired by an ad in ***Texas Monthly*** which advertised gold-painted chips for desk decorations for a hefty price. I have not made it past her home to see whether she really did it or not. While I do not recommend using cow chips to decorate a stucco wall, they are mentioned here to show that you can use just about anything to add interest to a stucco wall. Before you go out and do it, however, consider that with adequate humidity, they might be a wonderful growing medium for mushrooms.

Rustification or Rustication

Both of these words are commonly used in the stucco industry, but rustification is not found in Webster's New Explorer Dictionary. Maybe it is a Texas colloquial word (Texanese). In use, the connotation is that a building is made to look rural or rustic. Someone came up with the mistaken concept that rural things look rustic, and it stuck. Rustification is a general term, and it can include such things as installing a creaky wood floor in a Manhattan apartment, or installing a relatively-smooth stucco surface with intentional imperfections. So whether it is an actual word or not, in the vernacular, rustification means making a building look or seem rustic.

Rustication is a precise architectural term that relates to having smooth texture and rough texture on the same building. An example would be a rough stone wainscot with smooth stucco above. This has nothing to do with making a building look rustic. Some people use rustication to imply that a building should look rustic. If you see either of these words in a specification, it would be helpful if the person who wrote the specification defined exactly what was meant.

Quoins

Quoin is pronounced "coin." So when you hear that someone wants coins on their stucco building, they are not referring to installing dimes and quarters into the surface of the stucco, although this can be done. They are referring to installing corner decorations that are plant-ons that wrap around a corner. Usually they are installed with spaces between the quoins that are equal to the height of the quoins. Some quoins look horrible. Whoever selects the quoins to be used on a building should have some artistic ability and should consider how the quoins are going to look on the building. Prior to adding quoins to a building, it is appropriate to build the corner correctly before they are added. Otherwise, the corner may crack and with it the EPS quoins.

Chapter 22
Curing Stucco

In the early 1950s, Dr. Raymond E. Davis from the University of California at Berkeley, a renowned expert on Roman Cements, told my father that he had evidence that some of the cements that the Romans used 2,000 years ago were still gaining strength. While I do not know the specifics of that evidence, I suspect that it was the presence of calcium hydroxide and non-crystalline volcanic ash. To get the most out of the Roman Cement, maybe we should cure it for 2,000 years.

There are plasterers who slap mud on the wall and never make an attempt to ensure appropriate curing conditions, and as a result, problems occur within a few years. While I am not contending that stucco be cured for 2,000 years, it needs to be moist-cured much longer than most of it is currently being moist-cured.

Stucco is not like paint. Many paints cure by losing moisture. When the moisture is gone, they are cured. Stucco cures through a chemical reaction in which water combines chemically with the cement-like components of the stucco. During this process a number of chemical reactions are taking place.

In simplest terms, Portland cement powder is a mixture of different anhydrous calcium alumino-silicate compounds. Let's just take dicalcium silicate. When this compound is exposed to water, one of the calcium atoms is liberated as a calcium ion, and water is chemically attached to the remaining monocalcium silicate molecule. The calcium ion associates with a water molecule. Depending on the concentration of calcium ions in the water, it may remain in solution in its ionic state, or it may precipitate out as calcium hydroxide. If it remains in solution, and the exterior surface of the wall loses moisture, the calcium ions are carried to the surface with moisture that is moving to the surface. When they reach the surface, they combine with hydroxyl (OH^-) ions in the wall, and harden to form calcium hydroxide. At that point, the calcium hydroxide is water-soluble, and can be cleaned off with water. Over time, however, the calcium ions react with carbon dioxide in the air to form calcium carbonate. This is the same chemical that forms limestone. It bonds well to the exterior stucco surface. This is what is known as alkaline-earth efflorescence, or permanent efflorescence. The only practical way to remove it from the wall is with an acid wash.

Besides the dicalcium silicate, there are probably 100 other cementitious compounds in the Portland cement that are part of the stucco mud. Some react faster; and others slower. Most of them, if the stucco dehydrates before the chemical reaction takes place to hydrate the molecule, combine with carbon dioxide from the air to form carbonated calcium alumino-silicates. These molecules do not bond as well to other molecules as the hydrated calcium alumino-silicates, resulting in weak stucco.

If the water in the stucco mud remains in place, there is less likelihood of carbon dioxide entering the wall and producing the lower-grade carbonated molecules.

If the hydration process is interrupted by the loss of moisture, it is often difficult to get that hydration process started again. How long is the hydration process? A rule of thumb is that concrete reaches 90% of its ultimate strength at 28 days of age. That means if it is dehydrated at 28 days of age, it will never reach its ultimate strength. Modern building schedules do not allow 28 days for the stucco to cure.

Some masonry cements and stucco cements contain a pozzolan. This is a substance that has no cement properties on its own, but when in the presence of calcium ions and moisture will form hydrated cement molecules. This process is slow, so the stucco needs to remain moist for the reaction to occur. That reaction also ties up the calcium ions that are in the wall, and that can be carried to the surface to form permanent efflorescence. The longer a pozzolanic stucco is cured, the less likely it is to have permanent efflorescence.

Let's get away from the chemistry for a little bit and look at a wall that is freshly plastered. It tends to be a dark gray color. As time passes, some areas of the dark gray fade to a lighter gray. As more time passes more of the dark gray fades to a lighter gray. That dark gray color indicates that there is plenty of moisture present in the mud for chemical hydration to continue. That lighter gray color indicates that there is not enough moisture present for the chemical hydration to continue. That is a slight oversimplification, but pretty accurate.

If the wall dries out, it is difficult to get it wetted down again to re-create that dark gray stucco color. As a result, it is best to moisten the wall as often as necessary to prevent the dehydration. You are not trying to add water to the stucco, but rather to add as much water to the surface of the stucco as evaporates from the stucco.

The rate of hydration of the cement molecules is temperature-dependent. At lower temperatures, the rate slows and almost stops. If the wall freezes during the first 24 hours after applying the mud, the water in the wall turns to ice and expands. That is why ice floats on top of liquid water. When the water in the stucco freezes, it also expands and puts pressure on the stucco. This causes bonds to break, and pieces of stucco to come loose. As a result, stucco should not be allowed to freeze during that first critical 24 hours. After that first critical 24 hours, a portion of the water in the stucco has been chemically combined with the cement molecules. With that and with a minor loss of water from the stucco due to evaporation, there is some space within the stucco, so if the wall freezes, there is a possibility that no frost damage will occur. With each passing day, the wall can handle more freezing.

Some years ago I made up a series of stucco cubes. After 24 hours, I placed some of them in a deep freeze for a week. Afterwards I defrosted them and cured them in a conventional manner at about 70°F. All of the cubes ended up breaking at close to the same compressive strength when I broke them after 28 days of curing at 70°F. When I repeated the experiment and froze some of the cubes after 4 hours of curing, the results were much different. Several of the cubes that had been frozen broke in my hand before I could put them in the compressive test machine.

Because one does not want the stucco to freeze during that first 24 hours, it is common practice when cold weather is anticipated to put up plastic sheets and then heat the area between the plastic sheets and the stucco wall

This makes sense until one considers what can happen. If non-vented combustion is used, the carbon dioxide levels may build up in the enclosed areas and increase the carbonation of the cement molecules. If a few powerful units are used, portions of the enclosed area may still freeze while other areas may get overheated and become dehydrated.

To prevent situations like this, some people are tempted to use calcium chloride and destroy the metal lath in the stucco.

If the temperature is over 100° Fahrenheit or if a strong wind is blowing, the wall may dry out before chemical hydration can occur. To prevent the wall from drying out, curing compounds are sometimes applied to the stucco surface. Some of those curing compounds impair the bond with the next layer of stucco. If the finish coat will not stick to the brown coat, does it help to have a well-cured brown coat?

The first step in successfully curing stucco is to arrange your schedule so you do not have to apply stucco when freezing weather is anticipated or when strong drying conditions are likely to exist. If freezing weather is anticipated, plan the construction of the protection before you apply any stucco. After the stucco is applied, enclose it with protection and add gentle heat. A little heat over a long period of time is more effective than a lot of heat over a short period of time.

In hot weather, plan your day so you are never plastering in the sun. Come out early and start on the west walls so they will have a long curing period before the sun hits them. Then move to the southern walls. After the sun passes its zenith, move to the eastern walls, and finally finish on the northern walls.

So, how should a wall be moistened? Use a Hudson-type garden sprayer. Wait until the wall is thumbprint hard, so water will not erode it, and then lightly moisten the entire wall. If you use a garden hose, you may erode some of the stucco. Spray the wall whenever the wall starts appearing just a little bit light.

In Las Vegas, it is a common practice to hire retired couples to come past all freshly-plastered homes and moisten the walls several times per day.

There are curing aids available, but the only one that I trust is sodium meta-silicate. Any of the sodium silicates will work, but the sodium meta-silicate provides the most bang for the buck. Besides holding moisture in the wall, it will react with the calcium ions to plug the pores in the stucco, and often will restart the hydration process if it has been stopped.

Chapter 23
Texturing and Finishing Stucco

Texture can be applied as the third coat, using the same mud mix that was used for the brown coat. It may be a very similar mix, but with a different concentration of sand. It may be white Portland cement, hydrated lime, sand, and pigment. In other words, it may be a combined texture coat and color coat, or it may be a separate coat. It may be acrylic stucco with texture sand in it. If the texture coat was gray in color, a finish of stain or paint may be applied to provide color. The flatter the brown coat, the easier it is to apply a texture coat and make it look consistent throughout the job.

Cement-Like Texture Coats

When they start stuccoing, most people believe that a smooth stucco surface is the easiest finish to apply. After they have finished their first smooth job, they are convinced that it is the hardest finish to apply. It is so hard because every defect is highlighted. When the sun shines directly on the wall, it looks pretty good; but when the sun shines along the wall, the defects jump out.

At this point, you have learned to apply a Tuscany finish. I have heard it described as a relatively-smooth finish that is designed to look like it was applied by an applicator who does not know how to apply a smooth finish.

Seriously, textures on stuccos were designed to hide defects. That having been said, one of the hardest parts of applying a texture is leaving it alone after it has been applied. If there is a minor imperfection and you try to correct it, you will make it worse. A knock-down finish (commonly a splatter-sprayed finish that has been lightly hit with a trowel to remove the peaks) can become a Tuscany finish (a well worked finish with lots of variation in the surface and the color).

Another problem with any finish that most people do not realize until they have made a mistake is consistency. If you are applying a fan finish, as the day progresses and you get tired, the sweep of your trowel changes and the texture on one end of the wall may be larger than it is on the other end of the wall. The differences will show and be noticed. If you have two finish plasterers working on the same wall, if they have not trained to work together, ten years later when I am entering the building, I will point out that Trowel Man A did this section and Trowel Man B did that section. Other people may notice the differences as well, but are too polite to point them out to their companions. On the other hand, a large portion of the population is more interested in talking on cell phones and so never looks around at the beauty of the stucco you slaved to produce. At this point I have scratched two pages of text about people and their cell phones.

Prior to the development of the internet, most regions had a limited number of textures that were commonly used in that region. A texture might be called knock-down finish in one part of the country and Harry's special in another part of the country. Choosing a texture was easy since the choices were limited. Now, one can find documents that show over 30 different textures. A knock-down finish in Southern California may be different from a knock-down finish in North Carolina. Florida has variation in the names from the southern part to the panhandle of Florida. If a job specifies a Monterey finish, insist on a photograph of that finish that is approved by the specifier. Then is it a Monterey finish or a Monterrey finish? One is a California texture and the other is a Mexican texture that is commonly used in Texas. The Technical Services and Information Bureau, which is mentioned in the next paragraph, has a Monterrey finish that is nothing like the Monterrey finish that is used in South Texas. This book is printed with "print on demand" technology, and photographs do not reproduce well, so we will not attempt to develop our own set of names for different textures. Go to the web and do a search, and you will find numerous examples.

The Technical Services and Information Bureau, a Southern California stucco and plaster service bureau, has produced a website with photographs of 30 stucco and plaster textures. Each was taken from a distance of 4 feet and shows a 1 ft^2 area of the texture. Go to:

http://www.tsib.org/plaster.shtml to view their website. Another source of stucco textures is the Portland Cement Plaster/Stucco Manual, produced and published by the Portland Cement Association.

An easy texture to apply is to moisten the brown coat, spray mud on the wall with a hopper gun, and smooth it out with a trowel. The results will vary with the amount of sand in the mud, the length of time you wait before troweling, and the pressure you apply to the trowel. To vary the texture even more, spray a full coat of mud on the wall, wait 30 to 45 minutes, spray some more, and then hit it with a trowel. In South Texas we call these knock-down finishes.

In the older section of the little Alsatian town of Castroville, Texas, the measles finish is used. It is a relatively flat finish with rectangular bumps coming out of the wall at different angles. The bumps are produced by taking a full trowel of mud and applying it to one spot on the wall, and then smoothing the transition between the bump and the wall. It can hide some major imperfections in the flatness of the wall.

The Monterrey fan, also called the Spanish fan, is formed by starting low on the wall and working up in a straight line. With the trowel full of mud, it is arched from right to left (for a right-handed person). The next trowel-full of mud is applied about a trowel-length higher on the wall. Mud squeezes out below the trowel and adds a ridge to help define the fan. After reaching the top, the plasterer moves over and proceeds from the bottom to make another column of fans.

Practice is required. If at all possible, find a wall with a brown coat on it and apply the finish of your choice. After it is applied, scrape it off and apply it again. After a while, you will be able to concentrate on each stroke of the trowel and produce good results. Keep practicing until your arm learns the techniques and you do not have to think about each stroke of the trowel. If you can get a plasterer who knows a design to show you the application of the design and coach you in the application, it makes it much easier to learn.

No matter how carefully you mix cement-based colored stuccos, there will be variations in the color, based on different mixing techniques and times, and based on how the color coat cures and dries. If you want a color without any variations, find another surface. If you ended up with more variation than you wanted, a fog coat can be added. A fog coat consists of the fines from the color coat—the cement, hydrated lime, and the pigment. Mix it up so it can be sprayed through a Hudson-type sprayer. Strain it to remove all "oversized" particles. Two layers of pantyhose have been found to give satisfactory results. When applying the fog coat, apply a very thin layer to the wall, stand back and observe your handiwork, and then apply a little more where it is needed. Continue until the wall is colored the way you want it colored. If you want a wall that is one color on one end and gradually fades to another color, you can do it with two colors of fog coats. If your wall inadvertently does not have enough color variation, you can increase the color variation with a fog coat.

Carved Texture Coats

Back when the going rate for applying three-coat stucco, from the WRB to a Monterrey fan finish, was \$2.50 / ft^2, I wanted to apply a flagstone finish to a house I was building. A local plastering contractor who used the flagstone finish quoted me \$7.50 / ft^2, if I would provide him an acceptable brown coat. I'm cheap. In fact, I am so cheap that friends have wanted to know if I save leftover mud in a wheelbarrow to use on the next job. I went to the back of the house and started working. I knew that I needed a gray coat and a flagstone-colored coat. After troweling the coats on and carving mortar joints, I figured out that the texture I wanted could best be achieved by spraying both coats on with a dry-wall texture gun. The surface pattern could be varied by varying the air pressure, the fluidity of the stucco, the orifice of the gun, the distance and angle the gun was from the wall, and the speed at which the gun was moved. After achieving the surface texture I was happy with, I proceeded to carve on the surface. My favorite carving tool was a Nordmeyer Carving Tool. This was a piece of steel strap which I had bent back on itself and wrapped at the ends with duct tape to produce a comfortable handle. If I carved too early, the tool would gum up and the texture would tear. If I carved too late, I could not get the penetration I wanted. Carving when the surface was leather-hard was just right. The stucco being carved out would break into small particles and freely exit the carving tool without me having to stop to clean the tool.

Then the real test came. I selected a wall that was 40′ long and about 9′ tall (360 ft^2). It was on the back side of the house, so if I messed up, it would not show too badly. On a Saturday morning I started prepping for doing that wall. By the time a friend, Phil, who had never stuccoed, arrived at 9:00 am, I had the first batch of gray stucco mixed and was ready to apply it with the dry-wall texture gun. It was 2.7 ft^3 in volume, so applying it to the wall resulted in an average thickness of about 1/10″. We then mixed 4 ft^3 of the color coat and applied it in the same way. While Phil washed the equipment, I used a pencil to mark lines I wanted to carve. I did not do the entire wall, but did some along the entire length of the wall so I would have a tendency to carve "flagstone" of the same size along the entire length. When the stucco was leather-hard, Phil started carving where I had marked lines, and I started carving where I had not marked lines. At noon, we were in town eating lunch, and the wall was finished except for moist-curing for a few days. While eating lunch, we calculated the costs. If I had gone with the bid I had, it would have cost $2,620.00. My actual expenses were less than $35.00 for materials, and lunch for Phil and me which was

$8.00; we splurged that day because we had saved so much money. I had invested 4 hours into the job, after the 8 hours I spent teaching myself the technique; and Phil had invested 3 hours, plus he lived an hour away so he had a total of 2 hours driving. By adding all of those figures up, we were saving $151 per hour by doing the job ourselves. The rest of the house was done without having to learn the technique and with friends who lived closer, so we were saving about $320 per hour.

After the stucco had cured for a day, I used a sponge dipped in a mixture of water, white Portland, hydrated lime, and pigment to darken some of the "stones." By using several mixtures, I ended up with four different colored "stones" on the wall. Right in front of the house, I had a drip when I was adding extra color to one of the stones. When I was a kid, I could not color within the lines; and now as an adult, I still cannot color within the lines. I could have added extra pigment to the lower stone, but decided to leave it, showing that I have my limitations.

One wall was done on my wife's and my anniversary, so around the door and windows, I carved an anniversary message. The original plans called for trimming the windows and doors with cedar, but that has never happened.

Now, many years later, I can see problems with the wall. A number of the carved lines had a slight curve to them. Most flagstone has straight edges, rather than curved edges. All of my "stone" fit together well. Most flagstone do not fit together that well and have large areas of mortar showing periodically. Even with those problems, I have had people argue with me when I tell them that the house is a stucco house.

If you want to do a flagstone finish, you now have a road map to doing it; but if you want to do a finish that looks like cut stone, or brick, you have the information to develop the technique.

The same general technique is used if you have columns and you want to make them look like bark-covered tree trunks. With the brown coat, you can add extra stucco to simulate limbs coming out, or where limbs were cut off. Then spray with a darker stucco, and follow it with a lighter colored stucco.

Trowel smooth; and as it is drying, use nut picks to carve in the bark. In ***The Stucco Book—Creative Stuccoing***, we will delve further into applying finishes to stucco.

Acrylic Stucco

In simplest terms, acrylic stucco is an excellent grade of acrylic paint with texture sand added to it. The names of the textures are more standardized than with the cement-like textures, because the acrylic stuccos are made by manufacturers who have specified the names. There is some variation in names among various manufacturers, but they tend to go with

Smooth (acrylic paint),
Fine Sand,
Medium Sand, and
Heavy Sand.

The fine sand and medium sand textures can be applied by applying as smoothly as possible, or by imparting circular or other trowel marks to the surface. The heavy sand is usually applied with a circular motion to the trowel. This causes the largest pieces of sand to be dragged around by the trowel and produce a condition called "worm holes" in the surface. These are irregular indentations in the stucco that are the width of the sand particle and may be from 1/2" long to an inch or longer.

Painting Stucco

In Florida, I have observed more than one job where the contractor applied the cheapest exterior latex paint he could find; and to make the paint go further, he watered it down. This is not an acceptable finish for stucco. If one wants to paint a stucco surface to protect it and provide a uniform color, this is satisfactory, but a quality acrylic paint should be used, and it should not be diluted. Use an acrylic paint that is designed for stucco surfaces; otherwise, the alkali in the stucco may cause the paint to peel. Be sure that the paint will breathe. If it cannot and there is moisture in the stucco, blisters may form in the paint. Spraying paint onto a wall, especially one with texture, does not result in covering the entire surface of the wall. Rolling paint on a wall is a little better; but after the paint is rolled or sprayed onto a wall, it needs to be back-rolled. This is a process of rolling the paint roller, while full of paint, in the opposite direction from which it was first applied. After a wall is painted, it should be examined to see if all surfaces are covered, or whether there are areas around the texture particles where paint has not fully covered the wall.

If a wall has cracks, an elastomeric acrylic paint is an option. These paints are applied thicker than conventional paints, usually 10 mils or 12 mils dry thickness, and they are usually applied in two coats. If the cracks are wide enough so a credit card can fit into the crack, the crack should be filled before painting. To fill a crack, use a good grade of caulk or sealant and wipe up all of the material that does not end up in the crack. If any is left on the surface, it will change the surface texture. This change in texture, is sometimes more visible than the original crack. Again, roll and then back-roll each coat of elastomeric paint. Test the caulk or sealant out before you use it. With most caulks, after the surface has skimmed over, they may be painted without a problem. With some sealants, the paint will not adhere to them and will actually pull away from the sealant. Life is much easier if these are avoided.

A downside of elastomeric coatings is that they tend to mute the sharp texture that makes a stucco wall so attractive. The upside is that if there are cracks in the wall, the elastomeric coating will hide them and move as the cracks open and close, so it will continue to cover the cracks. Many elastomeric coatings provide 150% to 200% expansion. If a 1/8″ crack has been filled and painted over and then opens to 1/4″, the coating will stretch to accommodate the movement. If there is no crack and a crack opens to 1/8″, the coating will either crack or debond adjacent to the crack, because 150% of zero is still zero.

Adhered Concrete Masonry Veneer

Some years ago I referred to Adhered Concrete Masonry Veneer (ACMV) as fabulous texture on stucco. My friends in the Masonry Veneer Manufacturers' Association have not let me forget that comment and have repeatedly pointed out that while there are many similarities between installing ACMV and stucco, there are enough differences that the two should not be confused. Therefore, I conclude that if you add ACMV to your stucco wall, it is no longer a stucco wall but an ACMV wall and as such, is not covered further here. That having been said, in ***The Stucco Book—Creative Stuccoing***, I anticipate covering the use of ACMV.

Staining Stucco

Stucco can be stained in many ways. For a stain to penetrate and become a permanent part of the wall, the wall cannot be coated with a sealer. After the stucco has been stained and cured, a sealer can be added if desired.

If the wall contains polymer-modified cement (EIFS), it might as well be sealed, since the polymer-modified cement will not absorb the stain as readily as Portland-cement-based stucco. If you have plant-ons on the wall, determine whether there is any polymer-modified cement holding the plant-ons in place. It is a common practice to attach the plant-ons with polymer-modified cement and then add a texture coat on top of it. If this is done, and the texture coat does not cover all of the polymer-based cement (common with knock-down textures), then the plant-ons will stick out like sore thumbs wrapped in a white bandage.

Practice staining a sample panel before you mess up your house. Once you put stain on your house, you cannot take it off. You can add more so you have a darker color, but moving to a lighter color is very difficult.

A common method of staining a wall is to mix about 8 pounds of ferric nitrate or ferrous sulfate with one gallon of water. After it dissolves, strain out any particles, and use a Hudson-type sprayer to stain the walls. Better results are achieved by applying several light coats rather than one heavy coat. You can give extra coats to those areas where you want additional staining. Wait a week, examine the wall, and then determine whether you want to add more stain. Whatever you do, do not apply so much that the stain can run down the wall. Both of these stains produce an orange or rust color.

Stucco can also be stained with exterior paint pigments. Dilute one part pigment, by volume, in fifteen to twenty parts of water. Mix it up, place it in a Hudson-type sprayer, and apply it to the wall. At least every two minutes, swirl the sprayer so the pigment does not settle out. As with the above stains, you can apply several mist coats to obtain the color you want. If you would like your stucco home to look like it was built out of a large chunk of granite, you will have to wait for ***The Stucco Book—Creative Stuccoing***.

Chapter 24
Warranties

Traditionally, the manufacturers of stucco did not furnish warranties. The applicators warranted their work. Then when one-coat stucco came on the market, there were people who were leery about using a new product, so one of the early one-coat manufacturers added a ten-year warranty to his product in order to buy his way into the Florida market. Before long, every one-coat manufacturer who wanted to enter the Florida one-coat market needed to supply a warranty. In states close to Florida, there is a greater demand for warranties for one-coat work. While in states remote from Florida there is a periodic request for a warranty, it is nowhere near as prevalent as it is in Florida.

As the market has matured, the concept of warranties in Florida has spread to builders asking for a warranty for three-coat stucco.

Different manufacturers use warranties in different manners. The producers of acrylic finish coatings often use the warranties to sell their finish coats, and to ensure that their one-coat products are finish coated. Example: You use my one-coat stucco and you apply this finish coat, and you can receive a five-year warranty. If you apply the finish coat and then apply a sealer, you can receive a ten-year warranty.

Those manufacturers who do not have a line of top-coats have more trouble in getting adequate protection on their one-coat stucco.

Condominium owners, starting in Florida and spreading from there, started using warranties as a method to get a new paint job on their condos. Threaten a warranty lawsuit and then settle for some cash. Suddenly, warranties were written more carefully. Basically, a manufacturer can warrant the material in the bag, before the bag is opened. If he warrants anything else, he is warranting the application. Since the applicator is a separate contractor who can take shortcuts, warranting the applicator's work is a risk. Most one-coat warranties, unless issued jointly by the manufacturer and the applicator, are materials warranties. They basically say this product meets our published standards. The warranty may be for one year, or ten years, or fifteen years, but the coverage is only for the material before the bag is opened. The main reason a manufacturer should want to issue a warranty is that the warranty can limit what is covered. If a warranty is not issued, then common law prevails, and that can bring a few surprises.

What is actually covered in a warranty is often different from what the sales people state is covered in the warranty. Whether they are like congressmen and do not read documents they should be familiar with, or they are fluffing the product, I do not know; but I have repeatedly heard that Brand X gives a fifteen-year full warranty for materials and labor on their product. Their website states that they give a five-year warranty, and the copy of the warranty they actually issue is limited to the materials in the bag meeting their specifications.

A company I once worked with decided to settle the warranty issue once and for all. After one marathon session listening to attorneys, engineers, and salespeople discuss the needs for warranties, what should be covered, and what should not be covered, I nearly went to sleep. Since we were not getting any closer to a solution, I drafted the following warranty and slipped it in with a batch of warranties we were studying. Some saw the humor; some could not see any humor. You read it and decide. If you are not familiar with Nome, Alaska, the sun does not rise during the specified time.

Toobit Stucco Company
Unlimited Fifty-Year Warranty

We warrant our Toobit One-Coat Stucco against all failures for the life of the structure or fifty years, whichever comes first. If you have a stucco failure, for any reason, please apply in person for warranty adjustments. Our warranty covers the full replacement cost of the product less a $6.00 per bag processing fee. Labor, other materials, loss of time, damage to interior contents, hospital stays, etc., etc., etc., are not covered.

Failure to apply in person within 60 days of the time a competent inspector should have become aware of the problem is grounds to deny your just or unjust claim.

If any item was mixed with the Toobit One-Coat Stucco without express written authorization from the President of Toobit Stucco Company, this warranty is null and void. This includes water and sand.

Bring the appropriate documentation, as we cannot make warranty adjustments without proper documentation. As a minimum, bring sales tickets, invoices, and canceled checks, as well as all original packing material. Failure to bring all documentation is grounds to deny your just or unjust claim. Apply in person at our warranty claims office located 20 miles west of Nome, AK. Office hours are sunup to sundown from December 1 to February 28.

*Applicator Name*______________________________________

Signature ______________________________________

Date ___________

Toobit Stucco Company'

Signature ____

date ___________

One last thing about warranties, if they are printed on document paper and are suitable for framing, they are accepted much better than if they are printed on plain white paper, so if you are issuing warranties, invest $0.25 per sheet of paper to print them. It will lead to a lot fewer headaches.

Chapter 26
Helpful Hints

Keeping Equipment Clean

It is much easier to keep equipment clean, than to have to clean it after cement materials have hardened on it. When a break is taken for lunch, clean all of the equipment, regardless of whether it needs cleaning. Clean the mixer as if you were going to store it for a few weeks. At the end of the day, repeat the process.

When cleaning trowels, it is often difficult to get all of the stucco off the back side of the trowel. After wetting the trowel, place it on the ground with the handle up and move your feet back and forth on the top of the blade of the trowel to scrub the stucco off the trowel.

Prior to dumping mud into a wheelbarrow, wet the inside of the wheelbarrow. Mud will dump from the wheelbarrow easier, and it will make the wheelbarrow easier to clean.

Each little bit of hardened stucco is a collection point for more stucco to collect the next time you use the equipment.

If you clean your equipment after every four hours of use, it will provide a lot of service. If you do not, just buy the cheapest tools you can find.

If you are going to store your mixer with a gasoline engine for more than two weeks, run it until it runs out of gas. Some engines will gum up in as little as two weeks if gasoline is left in them.

Citric acid will attack hydrated cement, but not the steel that is used to make mixers, pumps, and wheelbarrows. Use a mix of about 25 grams of citric acid crystals and about 50 grams of water. Place an absorbent material around the offending hydrated cement and strongly moisten the absorbent materials with the citric acid solution. Keep it moist for a day, and check to see if it has softened. If it has not, continue the treatment for several more days.

Personal Protection

Stucco burns much more if your skin is moist. Keep your hands dry. If you get stucco on your skin, brush it off.

A long-sleeved cotton shirt keeps a person cooler than a short-sleeved shirt when working in the sun. The long sleeves can also keep stucco from getting on your skin.

Cotton gloves will not protect your hands if they get wet. They will just keep your hands moist, so stucco that gets onto the gloves can produce worse chemical burns.

After working with stucco all day, washing hands and arms with vinegar will reduce the severity of the chemical burns from the stucco. White vinegar does not have as strong an aroma as cider vinegar and is just as effective.

Stucco occasionally splatters. If you do not wear adequate eye protection, you should practice blinking before you get a dab of stucco in your eye. Stucco contains alkali material that can burn your eyes and sand that can abrade them.

If stucco gets into an eye, wash it thoroughly, and consider seeing an eye doctor.

Take a shower after the day is over, so you do not sleep with the alkali salts on your skin overnight.

Scheduling Work

One day I received a call from an architect who was involved with building a Federal Reserve building. The floor had been completed with stained concrete and was beautiful. Then the walls were being plastered. Someone forgot to adequately fasten down the plastic that had been installed to protect the floor. I was asked to come up with a solution to bring the stained concrete floor back to its original condition. I found that my schedule was too full to solve the problem. Seriously, there are some problems for which the only adequate solution is to not allow the problem to happen. Before a job starts, it should be planned out, every possible disaster should be listed, and a plan for preventing that disaster should be developed. In this case, the walls could have been done first, or the floor could have been protected with a better covering. That better covering could have included sealing and waxing the stained floor, as well as using a fabric to protect the floor. Additionally, someone should have been monitoring the protecting fabric to ensure that it remained in place.

Every component of a building weighs something. Every time a weight is added to a building, the added weight causes some component of the building to deflect. Sometimes those deflections are too small for us to measure. Sometimes those deflections can affect other components of the building. For example, if sheetrock is installed and then roofing shingles are placed on the roof, there may not be a problem. If roof tile are placed on the roof, and they weigh more than those roofing shingles, and thus will cause more deflection, they are likely to crack the sheet rock. Recently I read that one way to build "greener" was to reduce the framing lumber in a house. It was suggested that spacing studs at 24″ on centers was much "greener" than placing them at 16″ on centers. With the 24″ spacing even a lightweight roofing material may crack the sheetrock if it is installed before the roof is loaded. You do not care about sheetrock, so why am I talking about it? Two reasons—if roof loading will crack the sheetrock, it is likely to crack the stucco. Second, if the sheetrock is installed after the stucco is in place, nailing the sheetrock to the studs will cause the studs to move and will crack the stucco. My father taught me to install the roof before installing the sheetrock, and to install the sheetrock before installing the stucco. On more than one occasion I have started to explain this to an elderly builder, and they tell me that was one of the first things they learned when they started out. Some younger builders have never heard of this concept.

Substrates

Stucco cannot correct shoddy workmanship, nor should it be expected to hold a building together that has structural problems.

Some are otherwise-rational people make decisions as if they believe that stucco should be able to correct all problems on a construction site, including inadequate foundations.

Different substrates have different absorptions. The greater the absorption of a substrate, the more water it will suck out of the mud. The greater the need for the mud to carry more water, the more hydrated lime is needed to keep the mud workable as it carries that additional water.

Concrete and concrete products crack every 20′. Provision must be made to prevent those cracks or control their widths. If a concrete or unit masonry substrate cracks, stucco applied to that substrate will crack.

Stucco cannot bridge installed or inadvertent expansion joints in a building. An inadvertent expansion joint is a crack in a building that goes through a wall.

All concrete products shrink from the time they are produced until they are fully dry. Stucco shrinks about 0.14% or about 0.168 inches per ten feet.

All concrete products expand with increasing temperatures. Stucco expands at a rate of 5.9 to 7.0 times 10^{-6} / inches /inch / degree Fahrenheit. That translates to about 0.07 to 0.084 inches per 10 feet in length per 100° Fahrenheit. That is not much, but if the stucco is attached to something that expands at a different rate, bonds may be broken with each cooling and heating cycle.

Wood moves more than stucco, and each variety of wood moves at a different rate. Bond between stucco and wood cannot be considered permanent and usually results in the stucco cracking. The wood can be faced with felt paper to prevent cracks in the stucco. The felt paper needs to be covered with stucco lath or woven wire.

Every sheet of sheathing OSB and plywood includes a notice that a gap needs to be left between sheets. OSB and plywood are regularly gapped on roofs; it is not anywhere as common with sheathing. Un-gapped plywood or OSB can cause stucco to crack.

Stucco can wick moisture to wood, which can cause the wood to rot, so wood should be protected with a WRB, or it should be painted.

Metal expands much more than stucco with increased temperature, so stucco should not be applied directly to metal. Provide a separation consisting of felt paper or other substance.

Mixing Stucco

Don't put your hand, or anything else you do not want destroyed, in a mixer while it is running.

Mixing stucco over 5 minutes increases the air content and, the likelihood of plastic shrinkage cracks.

Mixing under 4 minutes may result in inadequately-mixed stucco and variable absorption on the wall.

When my father was 84-years-old, he said that he had yet to meet a person under 75 with enough sense to mix stucco correctly by hand.

Formulae Compromises

Higher hydrated lime concentration leads to:
Enhanced workability,
Higher cost,
Less bleed water,
Longer working time,
Lower ultimate compressive strength, and
Creamier stucco that will carry more sand.
Higher Portland cement concentration leads to:

Greater initial compressive strength,
Greater likelihood of plastic shrinkage cracks, and
Faster initial set.
Higher sand concentration leads to:
Lower strength,
Lower cost per cubic yard of prepared stucco,
Less likelihood of plastic shrinkage cracks,
Less workability, and
Less resistance to water penetration.
Higher air content leads to:
Greater workability of the mix,
Greater likelihood the stucco will crack from impact,
Less likelihood of cracks from plastic shrinkage,
Lower compressive strength, and
Lower flexural strength when air content is over 8%.

Cement

Different brands of Portland cement from the same region may behave in substantially-different manners.

Portland cement of the same brand, but from different plants, may behave in substantially-different manners.

Portland cement clinker is sold between plants within a region, so different batches of Portland cement from the same plant may behave in different manners.

Without considering pozzolanic reactions, the higher the Portland cement content, the greater the final compressive strength.

Lime

Air-entrained lime usually results in fewer cracks than non-air-entrained lime.

Other things being equal, Type S lime is less likely to burn skin than Type N lime.

Other things being equal, dolomitic lime is less likely to burn skin than high-calcium lime.

Sand

The coarser the sand, the less the workability and the less the likelihood of plastic shrinkage cracks.

Washed sand behaves more predictably than bank run sand.

Natural sand particles are often rounded. As a result, natural sands are usually more workable than manufactured sand (crushed) of the same gradation.

Sands with at least 5% sand retained on all sieves between 8-mesh and 100-mesh are more workable than sands with a narrower gradation.

Mortar Fat

ASTM does not have a listing for "mortar fats" which are used to replace lime in a mix. Several manufacturers of "mortar fat" state that mortar and stucco made with their product meet *ASTM* standards. Mortar fat is either bentonite clay, or an air entraining/plasticizing agent such as a ligno-sulfonate. Used in moderation and with accurate measurement, they are not bad. They usually work better when some hydrated lime is left in the mix. They are regularly misused.

Fiber

Fiberglass fibers protect against plastic shrinkage cracks (first 24 hours). Polypropylene fibers protect against plastic shrinkage cracks and reinforce the stucco for the life of the structure, but they have no value if exposed to a fire. Under certain conditions, fiberglass fibers can act as pozzolans and can react with calcium ions to form additional cement paste.

Fibrillated fibers (rough-edged fibers) provide a better mechanical bond to the stucco than monofilament fibers, but tend to ball up in the mixer.

Fibers can plug spray equipment.

Pigments

Stucco in which the pigments are well dispersed appears to be darker in color and provides a more consistent color than stuccos in which the pigments are not well dispersed.

Always add pigment to the mixer as soon as the initial sand and water are added to the mixer. The sand breaks down the clumps of pigment.

Careful volumetric measurements of some pigments result in errors of as much as 40%. Always weigh pigment for use in stucco.

When using any pigment, always measure all components.

When using pigment, add each component in the same order and as close as possible to the same time in the mixing cycle. After all components are in the mixer, mix for 5 minutes.

Never wipe your face while working with pigments.

When job-site-formulating colored stucco, assume that the Portland cement used will vary in color. Buy all of your cement from the same facility and at the same time. Arrange to take bags back rather than going in and buying a few more.

Premixed, pre-colored stucco varies in color from batch to batch.

Troweling changes the finished color of stucco.

Hadacall

I had to call this section something, so I called it Hadacall after a product that was on the market in the late 1940s. They had to call it something, so they called it Hadacall.

If you ever need to stop the set of stucco, add some sugar.

Never retemper your stucco with your soda.

Stucco transmits water. If you cannot place the stucco in front of a WRB, consider painting it with an elastomeric paint.

Elastomeric paint covers multiple cracks in stucco.

If everything else is equal, when you are plastering in the sun, the mud requires more water to remain workable than if you are plastering in the shade.

Probably one of the most important things to remember if the quality of your work is questioned is that there are no compressive strength requirements for stucco. There are for shotcrete, but not stucco. One of the reasons for this is that there are no accurate ways to test the compressive strength of in-place stucco. Inspectors love to use rebound hammers. They may be effective for testing the compressive strength of concrete, but most stuccos applied over lath and sheathing flex slightly which seriously impacts the results. Added to that, some pozzolanic stucco can be formed into a ball and bounced; this also seriously impacts the results. Penetration of power-actuated projectiles is no better at guesstimating the compressive strength of stucco, since the stucco is a thin layer and the projectile test is designed for thick sections of concrete. Cutting a section of the wall and inserting it in a compressive strength testing machine does not provide good results, since compressive strength is not only a function of the strength of the substance tested, but also a function of the height-to-width ratio of the test specimen.

After having been bored to tears at meetings when inspectors and others quote the Code and ASTM standards, I have found that many of those people have never read the Code or the ASTM standards they are quoting. Bringing a copy of the documents to the meeting and asking the verbose individual(s) to find the section he is quoting temporarily wins the argument, but makes enemies that you will face on future jobs. It is much better to furnish highlighted copies of the Code and the standards to the participants several days before the meeting to help educate them. On a number of occasions I have held classes on what different ASTM standards say and what they do not say.

There are people who build in areas that are not covered by any building code who believe that they do not have to comply with the requirements of a building code. This may be legally true, but never vary from the code until you fully understand the why behind the code language. If you follow this advice, you will find that you build better and you will find that it is easier to follow the code rather than deviate from the code.

Chapter 28
Stucco Resources

ASTM Standards

ASTM C 5 *Standard Specification for Quicklime for Structural Purposes*

ASTM C 91 *Standard Specification for Masonry Cement*

ASTM C 144 *Standard Specification for Aggregate for Masonry Mortar*

ASTM C 150 *Standard Specification for Portland Cement*

ASTM C 206 *Standard Specification for Finishing Hydrated Lime*

ASTM C 207 *Standard Specification for Hydrated Lime for Masonry Purposes*

ASTM C 260 *Standard Specification for Air-Entraining Admixtures for Concrete*

ASTM C 270 *Standard Specification for Mortar for Unit Masonry*

ASTM C 311 *Standard Test Methods for Sampling and Testing Fly Ash or Natural Pozzolans for Use in Portland-Cement Concrete*

ASTM C 585 *Standard Specification for Blended Hydraulic Cements*

ASTM C 618 *Standard Specification for Coal Fly Ash and Raw or Calcined Natural Pozzolan for Use in Concrete*

ASTM C 631 *Standard Specification for Bonding Compounds for Interior Plastering*

ASTM C 841 *Standard Specification for Installation of Interior Lathing and Furring*

ASTM C 847 *Standard Specification for Metal Lath*

ASTM C 897 *Standard Specification for Aggregate for Job-Mixed Portland Cement-Based Plasters*

ASTM C 926 *Standard Specification for Application of Portland Cement-Based Plaster*

ASTM C 932 *Standard Specification for Surface-Applied Bonding Agent for Exterior Plastering*

ASTM C 933 *Standard Specification for Welded Wire Lath*

ASTM C 979 *Standard Specification for Pigments for Integrally Colored Concrete*

ASTM C 1032 *Standard Specification for Woven Wire Plaster Base*

ASTM C 1063 *Standard Specification for Installation of Lathing and Furring to Receive Interior and Exterior Portland Cement-Based Plaster*

ASTM C 1116 *Standard Specification for Fiber-Reinforced Concrete and Shotcrete*

ASTM C 1157 *Standard Performance Specification for Hydraulic Cement*

ASTM C 1328 *Standard Specification for Plastic (Stucco) Cement*

ASTM C 1329 *Standard Specification for Mortar Cement*

ASTM C 1397 *Standard Practice for Application of Class PB Exterior Installation and Finish Systems*

ASTM C 1472 *Guide for Calculating Movement and Other Effects When Establishing Sealant Joint Width*

ASTM C 1489 *Standard Specification for Lime Putty for Structural Purposes*

ASTM C 1516 *Standard Practice for Application of Direct-Applied Exterior Finish Systems*

ASTM C 1535 *Standard Practice for Application of Exterior Insulation and Finish Systems Class PI*

ASTM C 1707 *Standard Specification of Specification for Pozzolanic Hydraulic Lime for Structural Purposes*

Other Resources

The **Portland Cement Association** has information concerning stucco and plastering on their website

http://www.cement.org/stucco/index.asp

Most of the major cement companies produce masonry cement, which is used for stuccoing. Some of them produce plastic (stucco) cement. Check their websites for information on stuccoing.

The **Northwest Wall and Ceiling Bureau** publishes a stucco manual which includes lots of details and a wonderful chart of textures. They also have considerable information on their website. http://www.nwcb.org/

Some stucco supply houses have a salesperson who is very knowledgeable about stucco. Others have salespeople who do not understand the product, but are willing to sell you what they have in their warehouses.

Remember, you can find just about anything on the web, and most of it has not seen peer review or an editor's pen. Some of it is good, but some of it is a waste of time. From the web I have learned that stucco is not breathable. I have learned stucco should never be used on straw structures. Go to the website http://apod.nasa.gov/apod/ap020401.html. Note the *nasa.gov*. This is a NASA website, and it indicates that the moon is made of green cheese. While this was done as an April Fools' joke, most erroneous information is posted up by people who do not know any better. Always test information to see if it makes sense. If it comes off the web, test it thrice.

Stucco-Related Papers by Nordmeyer

High Pozzolan Mortars and Stuccos *{14th International Symposium on Management and Use of Coal Combustion Products (CCPs), EPRI, Palo Alto, CA, 2001},*

Acid Resistance of High Fly Ash Mortars and Stuccos *{15th International Symposium on Management and Use of Coal Combustion Products (CCPs), EPRI, Palo Alto, CA, 2003},*

High Pozzolan Mortars and Stuccos *(Masonry: Opportunities for the 21st Century, STP 1432, ASTM, P.O. Box C700, West Conshohocken, PA 19428-2959, 2003},*

Degradation of One-Coat Stucco By Well-Meaning Professionals *{Walls and Ceilings Magazine, BNP Media, November 2005},*

Degradation of One-Coat Stucco By Well-Meaning Professionals-A Further Look *{Walls and Ceilings Magazine, BNP Media, February, 2006},*

Water-Repellent Performance in Pozzolanic and Traditional Mortars *(with Paul Taylor) {World of Coal Ash, 2007 Proceedings, American Coal Ash Associations, 15200 East Girard Ave, Suite 3050, Aurora, CO 50014, 2007},*

Type S Masonry Mortars Modified with Liquid Water Repellent *(with John H. Matthys) {10th North American Masonry Conference. The Masonry Society, 3970 Broadway, Suite 201-D, Boulder, CO 80304, 2007},*

Greening of Mortars with Pozzolans *(with Keith Bargaheiser) {Eleventh Symposium on Masonry, STP 1496, ASTM P.O. Box C700, West Conshohocken, PA 19428-2959, 2008},*

Variations in the Activity of Dry-Powder Water-Repellent Mortar Admixtures with Different Mortar Formulae *(with Pam Hall) {Eleventh Symposium on Masonry, STP 1496, ASTM P.O. Box C700, West Conshohocken, PA 19428-2959, 2008}.*

Index

About Nordmeyer, LLC

213 County Road 575
Castroville, TX 78009-2120
www.NordyBooks.com

Nordmeyer, LLC, is owned by the Nordmeyer family and was set up for:

- Their consulting business (stucco, mortar, pozzolans, green building, and building façade forensics),
- Their kayak guiding service (www.TxNatureKayaking.com), and
- Their writing and publishing business (www.NordyBooks.com).

For obvious reasons, the writing and publishing portion of the business is commonly referred to as Nordy Books. "Nordy Books" was set up to publish and market books that are written in whole or in part by Herb Nordmeyer. Those books fall into the following categories:

- Cancer,
- Construction-related,
- Devotions,
- Kayak-related, and
- Stories with life lessons.

Go to www.nordybooks.com for an update on which books are currently available and which ones will be published in the near future.

About Herb Nordmeyer

Herb is a leading stucco authority and has been published regularly in trade journals and ASTM publications. As a result, he is in demand as a speaker, an educator, an on-site consultant to solve construction issues, and an expert witness. Besides being admired for his knowledge, integrity, and candor, he is also known for a dry wit that is honed as sharply as is his mastery of the stucco industry.

Herb is an experienced wilderness kayak guide and a certified kayak instructor. He has been accused of using the same style in his classes as on wilderness expeditions—that of a loving grandfather teaching his beloved 16-year old grand child. He uses this style, even when his students are older than he is. Herb leads at least two nature kayaking trips per month. Some of those trips take 4 to 6 hours, and some require several days of primitive camping. The trips include

- Studying 10,000-year-old Indian art in caves that can only be reached by water,
- Drifting past Whooping Cranes as they feed on crabs,
- Smelling the fish-breath of dolphins as they surface close to a kayak,
- A quiet moon-lit paddle to watch 1.5 million Mexican free-tailed bats fly,
- Drifting down the wild and scenic Rio Grande with 1,200-foot-high cliffs on either side.

Herb is a prolific writer. Besides writing religious devotional pieces, he has developed several booklets that help people with cancer deal with spiritual issues. Besides this series of books on stucco, he is developing a series of books of true stories with life lessons. Expect a kayaking book within a few years.

Herb lives in Castroville, Texas with his wife Judy and the occasional stray cat or guinea fowl.

CPSIA information can be obtained
at www.ICGtesting.com
Printed in the USA
FFOW04n2057260214

9 780984 793617